CHEMICALS IN OUR FOOD:

What's REALLY on Your Plate?

KEVIN B DIBACCO

Staten House

CHEMICALS IN OUR FOOD,
Published by STATEN HOUSE

Copyright © 2024 by Kevin B. DiBacco
All rights reserved.

No part of this book may be reproduced in any form without written permission from the publisher or author, except as allowed by U.S. copyright law.

STATEN HOUSE is registered at
the U.S. Patent and Trademark Offices

Trade Paperback

Cover design © 2024 Yellow Dog Digital Studios.
All rights reserved.

Interior Formatting and Cover
Design Yellow Dog Digital Studios.

DISCLAIMER

No part of this publication may be reproduced in any form or by any means, including printing, scanning, photocopying, or otherwise, without the prior written permission of the copyright holder. The author has tried to present information that is as correct and concrete as possible. The author is not a medical doctor and does not write in any medical capacity. All medical decisions should be made under the guidance and care of your primary physician. The author will not be held liable for any injury or loss that is incurred to the reader through the application of the information here contained in this book. The author points out that the medical field is fast evolving with newer studies being done continuously, therefore the information in this book is only a researched collaboration of accurate information at the time of writing. With the ever-changing nature of the subjects included, the author hopes that the reader will be able to appreciate the content that has been covered in this book. While all attempts have been made to verify each piece of information provided in this publication, the author assumes no responsibility for any error, omission, or contrary interpretation of the subject present in this book. Please note that any help or advice given hereof is not a substitution for licensed medical advice. The reader accepts responsibility in the use of any information and takes advice given in this book at their own risk. If the reader is under medication supervision or has had complications with health-related risks, consult your primary care physician as soon as possible before taking any advice given in this book.

"The information and advice contained in this book are based upon the research and the personal and professional experiences of the author. They are not intended as a substitute for consulting with a healthcare professional. The publisher and author are not responsible for any adverse effects or consequences resulting from the use of any of the suggestions, preparations, or procedures discussed in this book. All matters pertaining to your physical health should be supervised by a healthcare professional."

CONTENTS

INTRODUCTION..1

CHAPTER ONE
INTRODUCTION TO FOOD CHEMISTRY - THE SECRET SCIENCE OF YOUR SNACKS...12

CHAPTER TWO
CHEMICAL ADDITIONS TO FOOD - THE HIDDEN INGREDIENTS ON YOUR PLATE 21

PART I
FOOD ADDITIVES AND PRESERVATIVES

PART II
PESTICIDES AND HERBICIDES IN AGRICULTURE

PART III
FOOD PROCESSING AND ITS CHEMICAL IMPACT

PART IV
HEALTH AND ENVIRONMENTAL CONSIDERATIONS

CHAPTER THREE
ALLERGENS AND SENSITIVITIES - WHEN FOOD FIGHTS BACK.39

CHAPTER FOUR
ENVIRONMENTAL CONTAMINANTS IN FOOD - UNINVITED GUESTS ON YOUR PLATE..49

CHAPTER FIVE
FOOD PACKAGING AND CHEMICAL MIGRATION 63

PART V
SPECIAL TOPICS IN FOOD CHEMISTRY

CHAPTER SIX
ORGANIC FOOD: MYTHS AND REALITIES 80

CHAPTER SEVEN
GENETICALLY MODIFIED ORGANISMS (GMOS) AND THEIR CHEMICAL PROFILES.. 89

CHAPTER EIGHT
THE IMPACT OF COOKING METHODS ON FOOD CHEMISTRY ... 99

PART VI
CONSUMER INFORMATION AND REGULATION

CHAPTER NINE
UNDERSTANDING FOOD LABELS AND CHEMICAL INGREDIENTS 116

CHAPTER TEN
REGULATORY FRAMEWORK FOR FOOD CHEMICALS 120

CHAPTER ELEVEN
EMERGING CONCERNS: NEW CHEMICALS IN OUR FOOD SUPPLY 125

CHAPTER TWELVE
PERSONALIZED NUTRITION - THE FUTURE OF EATING 132

GLOSSARY OF TERMS 150

INTRODUCTION

Hey there! Ever wonder what's really in that sandwich you're eating? Or why that packaged snack can sit on a shelf for months without going bad?

Welcome to the world of chemicals in our food. Don't worry – we're not talking about test tubes and lab coats. We're talking about the stuff that's in pretty much everything we eat, whether we know it or not.

Now, before you toss that apple or swear off pizza forever, let's get one thing straight: not all chemicals are bad. In fact, everything is made of chemicals – including you! Water? That's a chemical. The vitamin C in your orange? Also, a chemical. But here's the thing: some chemicals in our food might not be so great for us, and that's what this book is all about.

WHY SHOULD YOU CARE?

Okay, so why should you bother learning about chemicals in food? Well, for starters, what you eat affects your health. It's that simple. The food you put in your body can make you feel energized or sluggish, help you grow strong or make you sick.

And some of the chemicals in our food might be doing things to our bodies that we don't fully understand yet.

But it's not just about you. The chemicals used in growing and making our food can affect the environment too.

They can harm the soil, water, and even other animals. So, when we talk about chemicals in food, we are really talking about the health of our whole planet.

REAL MEDICAL STUDY: THE IMPACT OF FOOD ADDITIVES ON GUT HEALTH

To illustrate why understanding food chemicals is so important, let's look at a medical study. In 2019, researchers at Georgia State University conducted a study on the effects of common food additives on gut health. The study, published in the journal Nature, focused on two widely used emulsifiers:

carboxymethylcellulose (CMC)
polysorbate 80 (P80)

These emulsifiers are found in many processed foods, including ice cream, bread, and sauces. They help keep ingredients mixed and improve texture. Sounds harmless, right? Well, the study found some concerning results.

The researchers fed mice a diet having these emulsifiers at levels like what humans might consume. They discovered that the emulsifiers altered the mice's gut microbiome – the community of bacteria living in their intestines. This change in gut bacteria led to inflammation and metabolic syndrome, which includes conditions like obesity and diabetes.

But here's where it gets interesting. The researchers then transferred the altered gut bacteria from the mice that ate emulsifiers to mice with no gut bacteria of their own. Guess what? These mice also developed inflammation and metabolic syndrome, even though they never ate the emulsifiers directly!

This study suggests that some common food additives might be affecting our health in ways we didn't expect. It's a perfect example of why we need to pay attention to the chemicals in our food – even ones that seem harmless at first glance.

WHAT DO WE MEAN BY "CHEMICALS IN FOOD"?

When we say, "chemicals in food," we're talking about a bunch of different things:

1. **Natural chemicals:** These are the ones that are naturally part of the food. Like the caffeine in coffee or the sugar in an apple.
2. **Additives:** These are chemicals added to food on purpose. They might make food last longer, look better, or taste different.
3. **Pesticides:** These are used to kill bugs and other pests that damage crops.

4. **Contaminants:** These are chemicals that get into food by accident, like pollution from factories or packaging materials.

Some of these chemicals are fine and even good for us. Others? Not so much. The tricky part is figuring out which is which.

A BRIEF HISTORY OF FOOD AND CHEMICALS

People have been using chemicals to preserve and change their food for thousands of years. Ancient Egyptians used salt to keep meat from spoiling. Medieval Europeans used sulfur to keep wine fresh. These methods were important – they helped prevent hunger and made it possible to travel long distances with food.

But things really changed in the 20th century. Scientists figured out how to make all sorts of new chemicals, and the food industry started using them like crazy. Suddenly, food could last for months or even years. It could be any color of the rainbow. It could taste like anything you wanted.

At first, everyone thought this was amazing. No more worrying about food going bad! But then people started noticing some problems. Some of these new chemicals seemed to be making people sick. Others were harming the environment.

THE PROS AND CONS

Like most things in life, chemicals in food have their good and bad sides. Let's break it down:

PROS

- Food lasts longer, which means less waste
- We can have a wider variety of foods available year-round
- Some added chemicals, like vitamins, can make food more nutritious
- Food is often safer from harmful bacteria

CONS

- Some chemicals might have long-term health effects we don't fully understand yet
- The overuse of pesticides can harm the environment
- Some people have allergic reactions to certain food additives
- Natural nutrients might be lost when food is heavily processed

REAL-LIFE STORY: SARAH'S FOOD DETECTIVE JOURNEY

To understand how these pros and cons play out in real life, let's meet Sarah, a 35-year-old mom from California. Sarah never gave much thought to the chemicals in her food until her daughter, Emma, was born.

"When Emma started eating solid foods, I suddenly became hyper-aware of everything I was feeding her," Sarah recalls. "I started reading labels and was shocked by how many ingredients I couldn't even pronounce."

Sarah's curiosity led her down a rabbit hole of food research. She learned about pesticides, preservatives, and artificial colors. At first, she was overwhelmed and scared. "I felt like everything in the grocery store was potentially harmful," she says. "I considered going all-organic, but it was so expensive."

As Sarah dug deeper, she realized the issue wasn't black and white. She learned that some food additives, like certain preservatives, help prevent dangerous bacterial growth. She also discovered that organic foods aren't necessarily chemical-free – they just use several types of pesticides.

"I realized I needed to find a balance," Sarah explains. "I couldn't afford to buy everything organic, and I didn't want to make myself crazy worrying about every little thing."

So, Sarah developed a strategy. She focused on buying organic for the "Dirty Dozen" – fruits and vegetables known to have high pesticide residues. For other foods, she carefully read labels and avoided artificial colors and flavors when possible. She also started cooking more meals from scratch.

"It's not perfect, but I feel good about the choices I'm making," Sarah says. "And I'm teaching Emma to be aware of what she eats without being fearful."

Sarah's story shows how understanding food chemicals can empower us to make informed choices. It's not about avoiding all chemicals – that's impossible. It's about being aware and finding a balance that works for you and your family.

THE BIG DEBATES

There's a lot of argument about chemicals in food. Some people say we should go back to all-natural, unprocessed foods. Others say modern food technology is necessary to feed the world's growing population. The truth is probably somewhere in the middle.

One big debate is about genetically modified organisms, or GMOs. These are plants that have had their DNA changed to grow better or resist pests. Some people think they're a great solution to world hunger. Others worry they might have unexpected effects on our health or the environment.

Another hot topic is organic food. Organic farmers use fewer synthetic chemicals, but organic food is often more expensive. Is it worth it? That's something you'll have to decide for yourself.

WHY IT'S COMPLICATED

If some chemicals in food might be bad for us, why don't we just get rid of them all? Well, it's not that simple. Here's why:

1. We don't always know the long-term effects. Some chemicals might seem fine now, but cause problems years later.
2. Everyone's different. A chemical that's harmless to most people might make some people sick.
3. It's hard to study. There are so many chemicals in our food and environment that it's tough to figure out which ones are causing problems.
4. There's a lot of money involved. The food industry is huge, and changing how they do things can cost billions of dollars.
5. We need to feed a lot of people. Some argue that without modern food technology, we couldn't produce enough food for everyone.

WHAT CAN YOU DO?

Reading this, you might feel a bit overwhelmed. But don't worry – there's a lot you can do to make smart choices about the food you eat:

1. **Read labels:** Get in the habit of checking what's in your food.
2. **Eat a variety of foods:** Don't rely too much on any one type of food.
3. **Cook more:** When you make food yourself, you know exactly what's in it.
4. **Stay informed:** Keep learning about food and nutrition from reliable sources.

5. **Ask questions:** Don't be afraid to ask where your food comes from and how it's made.
6. **Wash your produce:** This can help remove surface pesticide residues.
7. **Consider organic for certain foods:** Prioritize organic for foods known to have high pesticide residues.
8. **Limit processed foods:** These often contain more additives and preservatives.
9. **Support local farmers:** Get to know the people who grow your food and their practices.
10. **Advocate for better food policies:** Make your voice heard on issues like food labeling and pesticide regulations.

THE FUTURE OF FOOD CHEMISTRY

As we wrap up this introduction, let's take a quick look at what the future might hold for food chemistry. Scientists are constantly working on new ways to make our food safer, more nutritious, and more sustainable. Here are a few exciting developments:

1. **Smart packaging:** Researchers are developing packaging that can detect when food is spoiling and change color to alert consumers.
2. **Personalized nutrition:** Advances in genetic testing might allow us to tailor our diets to our individual genetic makeup.

3. **Alternative proteins:** From lab-grown meat to insects, scientists are exploring new protein sources that could be more environmentally friendly.
4. **Nanotechnology:** Tiny particles could be used to improve food safety, enhance nutrient absorption, or even deliver medicines through food.
5. **Precision agriculture:** Using data and technology to apply pesticides and fertilizers more precisely, potentially reducing their overall use.

These innovations show that the world of food chemistry is always evolving. As we learn more about how chemicals in food affect our health and the environment, we can develop better ways to produce and consume food.

WHAT'S NEXT?

In this book, we're going to dive deep into the world of chemicals in our food. We'll look at what these chemicals are, where they come from, and what they might be doing to our bodies and our planet. We'll explore the latest scientific research, hear from experts, and learn about the laws that are supposed to keep our food safe.

But most importantly, we'll give you the information you need to make your own choices about what you eat. Because at the end of the day, that's what matters most – being able to decide for yourself what goes on your plate.

A snack (maybe an apple?), get comfortable, and let's start exploring the fascinating, complex, and sometimes crazy world of chemicals in our food. Trust us, you'll never look at your dinner the same way again!

Remember, knowledge is power when it comes to food. By understanding the chemicals in our food, we can make choices that are better for our health and the health of our planet. It's not about being perfect – it's about being informed and doing the best we can with the information we have.

So, are you ready to become a food detective? Let's get started on this exciting journey into the world of food chemistry!

CHAPTER ONE
INTRODUCTION TO FOOD CHEMISTRY – THE SECRET SCIENCE OF YOUR SNACKS

Have you ever wondered why bread rises, why apples turn brown when you cut them, or how cheese gets its flavor? Welcome to the world of food chemistry! Don't worry – we're not going to make you put on a lab coat or mix bubbling potions. Food chemistry is all about understanding the stuff that makes up our food and how it changes when we store, cook, or eat it.

WHAT IS FOOD CHEMISTRY, ANYWAY?

Food chemistry is like being a detective, but instead of solving crimes, you're figuring out what makes food tick. It's the science that explains why food looks, tastes, smells, and feels the way it does. Food chemists study the ingredients in our food – from the simplest foods like fruits and vegetables to the most processed snacks in the grocery store.

But why should you care about food chemistry? Well, for starters:

1. It helps you understand what you're eating

2. It can make you a better cook
3. It explains why some foods are good for you (and others, not so much)
4. It's behind a lot of cool food tricks and hacks

THE BUILDING BLOCKS OF FOOD

Just like everything else in the world, food is made up of chemicals. But don't freak out! Remember, water is a chemical too. When we talk about the chemicals in food, we're usually talking about these main categories:

1. **Carbohydrates:** These are your body's main source of energy. They include sugars, starches, and fiber.
2. **Proteins:** These are the building blocks of your body. They help you grow and repair tissues.
3. **Fats:** Despite their bad reputation, fats are important for your body. They help you absorb certain vitamins and keep you feeling full.
4. **Vitamins and Minerals:** These are like tiny helpers that keep your body running smoothly.
5. **Water:** Yep, good old H2O is a crucial part of most foods and our bodies.

Let's break these down a bit more:

CARBOHYDRATES: THE ENERGY MAKERS

Carbs are like the gasoline for your body's engine. They come in different forms:

- **Simple carbs** (like the sugar in candy) give you quick energy but don't last long.
- **Complex carbs** (like those in whole grains) give you energy that lasts longer.
- **Fiber** doesn't give you energy, but it helps keep your digestive system happy.

When you eat carbs, your body breaks them down into a simple sugar called glucose. This glucose travels through your blood to give energy to all parts of your body.

Fun Fact: When you eat more carbohydrates than your body needs right away, it stores the extra energy as fat. That's why eating too many carbs (especially simple ones) can lead to weight gain.

PROTEINS: THE BODY BUILDERS

Proteins are made up of smaller units called amino acids. Think of amino acids like Lego blocks – your body can put them together in separate ways to build whatever it needs.

Your body uses proteins to:

- Build muscles
- Make hormones (chemical messengers in your body)
- Create enzymes (which help chemical reactions happen in your body)
- Fight off infections

Some foods high in protein include meat, fish, eggs, beans, and nuts.

Cool Chemistry: When you cook an egg, the heat causes the proteins to change shape. This process is called denaturation, and it's why your egg goes from liquid to solid!

FATS: NOT THE BAD GUYS

Fats have gotten a bad rap, but they're super important. They:

- Help your body absorb certain vitamins (A, D, E, and K)
- Provide energy
- Help you feel full after eating
- Are necessary for brain health

There are distinct types of fats:

- **Saturated fats:** Found in animal products and some tropical oils. Eating too much can be bad for your heart.
- **Unsaturated fats:** Found in plants and fish. These are generally considered healthier.
- **Trans fats:** These are mostly human-caused and are not good for you. Many countries are banning them.

Chemistry in Action: Oil and water don't mix because of their chemical structure. Fats are hydrophobic, meaning they're scared of water!

VITAMINS AND MINERALS: THE MICRONUTRIENT SQUAD

Vitamins and minerals are needed in tiny amounts, but they play huge roles in keeping you healthy.

Vitamins: These are organic compounds (meaning they hold carbon) that your body needs to function properly. There are two main types:

1. **Water-soluble vitamins** (like vitamin C and B vitamins): Your body doesn't store these, so you need to eat them regularly.
2. **Fat-soluble vitamins** (A, D, E, and K): Your body can store these in fat, so you don't need them every day.

Minerals: These are inorganic elements that your body needs. Some examples are calcium (for strong bones), iron (for healthy blood), and sodium (for nerve function).

WATER: THE UNSUNG HERO

Water might seem boring, but it's crucial for life. In food chemistry, water is important because:

- It's a solvent, meaning it can dissolve other substances
- It affects the texture of food
- It's necessary for many chemical reactions in cooking

Did You Know? The way water behaves is behind many cooking techniques. For example, when you boil water, the bubbles are water vapor pushing up through the liquid water.

CHEMICAL REACTIONS IN FOOD

Now that we know the main players let's talk about how they interact. Chemical reactions are happening in your food all the time – when you cook, when food spoils, and even when you digest it.

Here are some common types of reactions in food:

1. **Maillard Reaction:** This is what makes food brown and develops flavor when you cook it. It happens when proteins and sugars get heated together. Think of the crust on bread or the sear on a steak.
2. **Caramelization:** This is what happens when you heat sugar. It turns brown and gets a rich, sweet flavor. This is how we make caramel!
3. **Oxidation:** This is when food is exposed to oxygen. Its why apples turn brown when cut or why oil can go rancid.
4. **Fermentation:** This is when microorganisms (like yeast or bacteria) break down sugars. It's how we make bread, cheese, yogurt, and even some drinks like beer and kombucha.

FOOD CHEMISTRY IN THE KITCHEN

You don't need a lab to be a food chemist – your kitchen is a chemistry lab! Here are some everyday examples of food chemistry:

1. **Baking a cake:** When you mix baking soda (a base) with an acidic ingredient like buttermilk, it creates bubbles of carbon dioxide. These bubbles make your cake rise.
2. **Making salad dressing:** Oil and vinegar don't mix naturally. But if you add an emulsifier like mustard, it helps them stay mixed.
3. **Cooking meat:** The heat causes proteins in the meat to change shape (remember denaturation?). This is why meat changes color and texture when cooked.
4. **Ripening fruit:** As fruit ripens, enzymes break down complex carbohydrates into simpler sugars. That's why ripe fruit is sweeter!

THE DARK SIDE OF FOOD CHEMISTRY

While food chemistry can create delicious and nutritious foods, it can also be used in ways that might not be so great for our health:

1. **Artificial additives:** Some of these can cause allergic reactions or other health issues in some people.
2. **Ultra-processed foods:** These often use food chemistry tricks to make them hyper-palatable (super tasty), which can lead to overeating.

3. **Loss of nutrients:** Some processing techniques can strip foods of their natural nutrients.
4. **Hidden sugars and fats:** Food scientists can use chemistry to hide less healthy ingredients in our food.

FOOD CHEMISTRY AND YOUR HEALTH

Understanding food chemistry can help you make healthier choices:

1. **Reading labels:** Knowing the basic chemistry of food can help you understand what's really in your food.
2. **Cooking methods:** Different cooking methods can affect the nutritional value of food. For example, boiling vegetables can cause water-soluble vitamins to leach out into the water.
3. **Food combinations:** Some nutrients work better together. For instance, vitamin C helps your body absorb iron from plant sources.
4. **Food sizes:** Understanding energy density (calories per gram) can help you manage part sizes better.

THE FUTURE OF FOOD CHEMISTRY

Food chemistry isn't just about understanding what's in our food now – it's also about creating the food of the future. Scientists are working on some cool stuff:

1. **Lab-grown meat:** Using cells from animals to grow meat in a lab, without having to raise and slaughter animals.

2. **3D printed food:** Imagine downloading a recipe and "printing" out your dinner!
3. **Personalized nutrition:** Using your genetic information to create a diet perfectly tailored to your body's needs.
4. **Sustainable ingredients:** Finding new sources of protein and other nutrients that are better for the environment.

WRAPPING IT UP

Food chemistry might seem complicated, but it's really all about understanding the building blocks of what we eat and how they interact. By learning a bit about food chemistry, you can become a better cook, a smarter shopper, and maybe even a healthier eater.

Remember, every time you cook a meal, mix a smoothie, or even just bite into an apple, you're experiencing food chemistry in action. So, the next time someone tells you that science isn't practical, just invite them over for dinner – you've got some tasty chemistry experiments to show them!

CHAPTER TWO
CHEMICAL ADDITIONS TO FOOD - THE HIDDEN INGREDIENTS ON YOUR PLATE

INTRODUCTION

Hey there, food detectives! Ready to uncover the secret world of chemical additions in our food? Buckle up, because we're about to take a deep dive into the stuff that's added to our grub - some of it good, some not so much, and some downright mysterious. We're talking about additives, preservatives, pesticides, herbicides, and all the way food processing can change what ends up on our plates.

Now, don't freak out! Not all chemical additions are bad. Some help keep our food safe, make it last longer, or even add nutrients. But others? Well, they might be doing more harm than good. Our job is to learn about these additions so we can make smart choices about what we eat. In this chapter, we are going to look at:

1. Food Additives and Preservatives
2. Pesticides and Herbicides in Agriculture
3. How Food Processing Affects Our Food

PART I

FOOD ADDITIVES AND PRESERVATIVES

WHAT ARE FOOD ADDITIVES?

Food additives are substances added to food to keep it fresh, improve its taste, texture, appearance, or nutritional value. They've been used for centuries - think of salt for preserving meat or vinegar for pickling vegetables. But in the last century, we've seen an explosion of new, synthetic additives.

TYPES OF ADDITIVES

1. **Preservatives:** These keep food from spoiling.
2. **Colorants:** They make food look more appealing.
3. **Flavor enhancers:** These make food taste better.
4. **Texturizers:** They improve the texture or consistency of food.
5. **Nutritional additives:** These add vitamins or minerals.

Let's break these down and look at some common examples:

PRESERVATIVES

- **Sodium Benzoate (E211):** Found in sodas, fruit juices, and jams. It prevents the growth of bacteria and fungi.
- **Potassium Sorbate (E202):** Often used in wines, cheese, and baked goods. It stops mold and yeast growth.
- **Nitrites and Nitrates (E249-E252):** Used in processed meats like bacon and hot dogs. They prevent bacterial growth but have been linked to cancer risks.
- **Sulfites (E220-E228):** Found in dried fruits, wines, and some processed foods. They prevent browning and bacterial growth but can trigger asthma in some people.

Real-world example: Check out the ingredient list on a package of sliced bread. You might see calcium propionate or potassium sorbate - both are preservatives that keep the bread from getting moldy.

COLORANTS

- **Tartrazine (E102):** A yellow dye used in candies, cereals, and some pickles. It's banned in some countries due to potential health risks.
- **Carmine (E120):** A red dye made from crushed insects. It's used in yogurts, candies, and some meat products.
- **Caramel Color (E150):** Used in cola drinks, soy sauce, and some breads. In high doses, it may have carcinogenic effects.

Real-world example: Ever wonder why your maraschino cherries are so red? They're often colored with Red 40 (Allura Red AC).

FLAVOR ENHANCERS

- **Monosodium Glutamate (MSG) (E621):** Commonly used in Chinese food, canned vegetables, and soups. It enhances savory flavors but may cause headaches in some people.
- **Disodium Inosinate (E631):** Often used with MSG in instant noodles, snack foods, and canned fish.
- **Artificial Sweeteners:** Like aspartame (E951) in diet sodas or sucralose (E955) in low-calorie desserts.

Real-world example: Many ranch dressings use MSG to boost their savory flavor.

TEXTURIZERS

- **Xanthan Gum (E415):** Used in salad dressings, sauces, and gluten-free products to thicken and stabilize.
- **Carrageenan (E407):** Derived from seaweed, it's used in dairy products like ice cream to prevent separation.
- **Lecithin (E322):** Found in chocolate, margarine, and baked goods. It helps mix ingredients that usually don't mix well, incompatible.

Real-world example: Ever noticed how your store-bought ice cream doesn't get icy, even after being in the freezer for months? Thank texturizers like carrageenan for that smooth consistency.

NUTRITIONAL ADDITIVES

- **Ascorbic Acid (Vitamin C) (E300):** Added to fruit juices and cereals to boost nutritional value and act as an antioxidant.
- **Calcium Carbonate:** Added to plant-based milks and some fruit juices for calcium fortification.
- **Folic Acid:** Added to flour and cereals to prevent birth defects.

Real-world example: Most milk in the U.S. is fortified with Vitamin D to help prevent rickets and osteoporosis.

THE CONTROVERSY AROUND ADDITIVES

While many additives are considered safe, some have been linked to health concerns:

- **Artificial Colors:** Some studies suggest they may contribute to hyperactivity in children.
- **Sodium Nitrite:** When heated to hot temperatures, it can form potentially cancer-causing compounds.
- **BHA and BHT:** These preservatives have shown mixed results in studies, with some suggesting they could be carcinogenic.
- **Artificial Sweeteners:** While generally considered safe, some people report side effects like headaches or digestive issues.

Real-world example: In 2021, Skittles faced a lawsuit over their use of titanium dioxide, a color additive that's banned in some countries due to potential health risks.

PART II
PESTICIDES AND HERBICIDES IN AGRICULTURE

WHAT ARE PESTICIDES AND HERBICIDES?

Pesticides are chemicals used to kill pests that damage crops. Herbicides are a type of pesticide specifically used to kill weeds. While they help increase crop yields, some can leave residues on our food.

COMMON PESTICIDES AND THEIR USES

- **Glyphosate:** A widely used herbicide, often found on corn, soybeans, and wheat.
- **Chlorpyrifos:** An insecticide used on fruits, vegetables, and nuts. It's been banned for residential use due to health concerns.
- **Atrazine:** An herbicide commonly used on corn. It's been linked to hormonal disruptions in animals.
- **Malathion:** An insecticide used on fruits, vegetables, and landscaping. It can be toxic to bees and other beneficial insects.

Real-world example: Strawberries often top the "Dirty Dozen" list of produce with the most pesticide residues. A single sample has been found to hold up to 23 different pesticides!

ORGANIC VS. CONVENTIONAL FARMING

Organic farming uses natural pest control methods and organic fertilizers instead of synthetic pesticides and herbicides. However, organic doesn't mean pesticide-free - organic farmers can use certain natural pesticides.

Conventional farming often relies more heavily on synthetic pesticides and herbicides. While these chemicals are regulated, some residues can remain on food.

Real-world example: A conventional apple might be treated with up to 16 different pesticides, while an organic apple might be treated with natural pesticides like neem oil or diatomaceous earth.

THE IMPACT OF PESTICIDES ON HEALTH AND ENVIRONMENT

- **Health Concerns:** Some pesticides have been linked to cancer, hormone disruption, and neurological problems.
- **Environmental Impact:** Pesticides can harm beneficial insects, contaminate water sources, and contribute to soil degradation.
- **Resistance:** Overuse of pesticides can lead to resistant "superbugs" and "superweeds."

Real-world example: The monarch butterfly population has declined by 80% in the last two decades, partly due to herbicides killing milkweed, their primary food source.

REDUCING PESTICIDE EXPOSURE

- Wash and peel fruits and vegetables
- Buy organic, when possible, especially for the "Dirty Dozen"
- Grow your own produce
- Support integrated pest management practice

PART III
FOOD PROCESSING AND ITS CHEMICAL IMPACT

WHAT IS FOOD PROCESSING?

Food processing involves transforming raw ingredients into food products. This can be as simple as cutting and washing vegetables or as complex as creating a frozen TV dinner.

TYPES OF FOOD PROCESSING

1. **Minimal Processing:** Washing, cutting, packaging
2. **Basic Processing:** Cooking, freezing, canning
3. **Moderate Processing:** Adding preservatives, flavoring
4. **Highly Processed:** Creating ready-to-eat meals, snack foods

HOW PROCESSING AFFECTS OUR FOOD

Processing can affect food in several ways:

- **Nutrient Loss:** Heat, light, and oxygen exposure during processing can destroy some nutrients.
- **Nutrient Enhancement:** Some nutrients may be added back in (fortification) or become more available.
- **Chemical Changes:** Cooking can create new compounds, some beneficial (like antioxidants in tomatoes), others potentially harmful (like acrylamide in fried foods).
- **Texture Changes:** Processing can dramatically alter food texture, often through the addition of emulsifiers, stabilizers, or other additives.

Let's look at some common processed foods and how they're affected:

BREAKFAST CEREALS

- **Processing:** Grains are milled, mixed with additives, shaped, and toasted.
- **Chemical Additions:** Often fortified with vitamins and minerals. May have added sugars, artificial colors, and preservatives.
- **Impact:** Can be a reliable source of fiber and added nutrients, but often high in sugar and low in natural nutrients.

Real-world example: A popular cereal like Froot Loops has over 10 grams of sugar per serving, artificial colors, and BHT as a preservative.

BREAD

- **Processing:** Flour is mixed with yeast, water, and other ingredients, then baked.
- **Chemical Additions:** May hold dough conditioners, preservatives, and added vitamins.
- **Impact:** Commercial bread often lasts longer than homemade but may have more additives.

Real-world example: Many commercial breads contain azodicarbonamide, a dough conditioner that's also used to make yoga mats!

CHEESE

- **Processing:** Milk is cultured, enzymes are added to separate curds and whey, then it's aged.
- **Chemical Additions:** May have colorants (like annatto), anti-caking agents, or mold inhibitors.
- **Impact:** Processed cheese products often have emulsifiers and extra salt.

Real-world example: American cheese isn't legally cheese - it's a "pasteurized prepared cheese product" due to its high additive content!

YOGURT

- **Processing:** Milk is fermented with bacterial cultures.
- **Chemical Additions:** Often has added sugars, artificial flavors, and stabilizers like pectin or gelatin.
- **Impact:** While yogurt is generally healthy, flavored varieties can be high in added sugars.

Real-world example: Some fruit-flavored yogurts contain more sugar per ounce than ice cream!

DELI MEATS

- **Processing:** Meat is cured, sometimes smoked, and often contains added water and phosphates.
- **Chemical Additions:** Usually has nitrites or nitrates as preservatives.
- **Impact:** Convenient but often high in sodium and preservatives.

Real-world example: A single slice of bologna can have up to 300 mg (about the weight of ten grains of rice) of sodium - that's 13% of your daily recommended intake!

SOFT DRINKS

- **Processing:** Water is carbonated and mixed with flavorings, sweeteners, and other additives.
- **Chemical Additions:** Has artificial sweeteners or high fructose corn syrup, flavorings, and often caffeine.
- **Impact:** High in empty calories, may contribute to obesity and tooth decay.

Real-world example: A 12 oz (about 354.88 ml) can of cola holds about 39 grams of sugar - that's nearly 10 teaspoons!

FROZEN DINNERS

- **Processing:** Ingredients are pre-cooked, assembled, and flash frozen.
- **Chemical Additions:** Often high in sodium, may contain preservatives, flavor enhancers, and artificial colors.
- **Impact:** Convenient but often less nutritious than home-cooked meals.

Real-world example: Some frozen dinners have up to 1,500 mg (about half the weight of a penny) of sodium - that's an entire day's worth in one meal!

THE ULTRA-PROCESSING PROBLEM

Ultra-processed foods are industrial formulations made mostly from substances extracted from foods (like oils, sugars, starches), derived from food constituents (like hydrogenated fats, modified starch), or synthesized in laboratories (like flavor enhancers, colors).

Examples of ultra-processed foods:

- Soft drinks
- Packaged snacks
- Chicken Nuggets
- Instant noodles
- Many breakfasts cereals

Concerns about ultra-processed foods:

- Often high in added sugars, unhealthy fats, and sodium
- Low in fiber and essential nutrients
- May hold additives with potential health risks
- Linked to obesity, heart disease, and other health problems

Real-world example: A study found that 60% of the calories in the average American diet come from ultra-processed foods!

MAKING INFORMED CHOICES

While it's nearly impossible to avoid all processed foods, here are some tips to make healthier choices:

1. **Read Labels:** Look for products with fewer ingredients, especially ones you can pronounce.
2. **Choose Whole Foods:** opt for fresh fruits, vegetables, whole grains, and lean proteins.
3. **Cook More:** Homemade meals give you control over ingredients.
4. **Be Wary of Health Claims:** "Low-fat" or "sugar-free" doesn't always mean healthy.
5. **Moderation is Key:** It's okay to enjoy processed foods sometimes, just don't make them the bulk of your diet.

Chemical additions to our food - whether they're additives, pesticides, or the result of processing - are a complex topic. While many of these additions serve important purposes, like keeping our food safe and accessible, others may pose risks to our health and the environment. By understanding what's in our food and how it gets there, we can make more informed choices about what we eat. Remember, you have the power to decide what goes on your plate. So, the next time you're at the grocery store, take a moment to think about the journey your food has taken to get there - and what might have been added along the way.

Eating doesn't have to be complicated. By choosing a variety of whole, minimally processed foods most of the time, you're already making a great start towards a healthier diet. And hey, if you enjoy a bag of chips or a soda now and then, that's okay too. The key is balance and being aware of what you're putting into your body.

So go forth, young food scientists! Armed with this knowledge, you're ready to navigate the wild world of modern food. Your taste buds - and your body - will thank you.

PART IV
HEALTH AND ENVIRONMENTAL CONSIDERATIONS

CHAPTER THREE
ALLERGENS AND SENSITIVITIES - WHEN FOOD FIGHTS BACK

Hey there, food detectives! Today we're diving deep into the world of food allergies and sensitivities. It's like your body becomes a superhero, fighting off what it thinks are dangerous invaders - except sometimes, those "invaders" are just harmless foods. Let's unravel this mystery!

WHAT ARE FOOD ALLERGIES?

A food allergy is like your body's security system going into overdrive. Your immune system mistakenly names a certain food as a threat and launches an attack. This can cause symptoms ranging from mild (like a little itching) to severe (like difficulty breathing).

THE SCIENCE BEHIND ALLERGIES

When you have a food allergy, your immune system produces antibodies called Immunoglobulin E (IgE) in response to the food. These antibodies trigger the release of chemicals like histamine, which cause allergy symptoms.

The Big 9 Food Allergens:

1. Milk
2. Eggs
3. Peanuts
4. Tree nuts (like almonds, walnuts, and cashews)
5. Fish
6. Shellfish (like shrimp and crab)
7. Soy
8. Wheat
9. Sesame (newly added in 2021)

Let's break these down in detail:

1. MILK ALLERGY

- Most common in infants and young children
- Usually outgrown by adulthood (about 80% outgrow it by age 16)
- **Symptoms:** Hives, vomiting, digestive problems
- **Hidden sources:** Casein (milk protein) in some canned tuna, sausages, and non-dairy products

Fun Fact: Some people with milk allergies can tolerate goat's or sheep's milk, as the proteins are slightly different.

2. EGG ALLERGY

- Second most common food allergy in children
- Most kids outgrow it by age 16
- **Symptoms:** Skin reactions, digestive issues, rarely anaphylaxis

- **Hidden sources:** Some vaccines (like flu shots), hair care products, and even some wines use egg in processing Did You Know? Some people are allergic only to egg whites or only to egg yolks!

3. PEANUT ALLERGY

- Often causes severe reactions
- Usually lasts a lifetime (only about 20% of kids outgrow it)
- **Symptoms:** Can range from mild to severe, including anaphylaxis
- **Hidden sources:** Some vegetarian meat substitutes, pet food, and even some sunscreens!

Peanut Allergy Breakthrough: In 2020, the FDA approved Palforzia, the first drug for treating peanut allergies in children. It works by gradually exposing kids to tiny amounts of peanut protein to build tolerance.

4. TREE NUT ALLERGY

- Includes almonds, walnuts, cashews, pistachios, and more
- Often co-occurs with peanut allergy (but peanuts are legumes, not nuts!)
- **Symptoms:** Like peanut allergy, can be severe
- **Hidden sources:** Pesto sauce (often holds pine nuts), some non-dairy milk alternatives

5. FISH ALLERGY

- More common in adults than children
- Usually lasts a lifetime
- **Symptoms:** Hives, nausea, headaches, and in severe cases, anaphylaxis
- **Hidden sources:** Caesar salad dressing (often contains anchovies), Worcestershire sauce

Fishy Fact: Some people are allergic to only certain types of fish. For example, you might be allergic to salmon but able to eat cod safely.

6. SHELLFISH ALLERGY

- Includes crustaceans (like shrimp and crab) and mollusks (like clams and squid)
- More common in adults
- **Symptoms:** Can be severe, including anaphylaxis
- **Hidden sources:** Some Asian cuisines use shrimp paste as a flavoring

Did You Know? Shellfish allergies are more common in countries with high seafood consumption, like Japan and Southeast Asian countries.

7. SOY ALLERGY

- More common in infants and young children
- Often outgrown by adulthood
- **Symptoms:** Usually mild, like hives or itching

- **Hidden sources:** Many processed foods, including some canned broths and bouillon cubes

Soy Surprise: Soy is one of the "Big 8" allergens, but severe reactions are relatively rare compared to other major allergens.

8. WHEAT ALLERGY

- Different from celiac disease (which is an autoimmune reaction to gluten)
- Often outgrown by adulthood
- **Symptoms:** Hives, digestive issues, in rare cases anaphylaxis
- **Hidden sources:** Soy sauce, some ice creams, even some makeup and play dough!

Wheat vs. Gluten: Remember, wheat allergy is not the same as gluten sensitivity or celiac disease. Someone with a wheat allergy might be able to eat other grains containing gluten, like barley or rye.

9. SESAME ALLERGY

- Newly recognized as a major allergen in the U.S. as of 2021
- Becoming more common, possibly due to increased consumption
- **Symptoms:** Can range from mild to severe
- **Hidden sources:** Some spice blends, vegetable oils, and protein bars

Sesame Surge: A study published in JAMA Network Open in 2019 estimated that over 1.5 million Americans may have a sesame allergy, which is why it was added to the major allergen list.

FOOD SENSITIVITIES: THE SNEAKY COUSINS OF ALLERGIES

Food sensitivities are different from allergies. They don't involve the immune system in the same way, but they can still cause uncomfortable symptoms. These reactions are often delayed, making them tricky to name.

COMMON FOOD SENSITIVITIES

1. LACTOSE INTOLERANCE

- Inability to digest lactose, the sugar in milk
- **Symptoms:** Bloating, gas, diarrhea
- **Management:** Lactase enzymes, lactose-free dairy products

2. GLUTEN SENSITIVITY

- Different from celiac disease or wheat allergy
- **Symptoms:** Digestive issues, headaches, fatigue
- **Management:** Gluten-free diet

3. MSG SENSITIVITY

- Reaction to monosodium glutamate, a flavor enhancer
- **Symptoms:** Headaches, flushing, sweating

- **Hidden sources:** Some Chinese food, canned vegetables, soups

4. SULFITE SENSITIVITY

- Reaction to sulfites used as preservatives
- **Symptoms:** Asthma symptoms, hives
- **Hidden sources:** Dried fruits, wine, some medications

5. FODMAP SENSITIVITY

- Reaction to certain types of carbohydrates
- Common in people with IBS (irritable bowel syndrome)
- **Symptoms:** Bloating, gas, abdominal pain
- **Management:** Low FODMAP diet under guidance of a dietitian

DEALING WITH FOOD ALLERGIES AND SENSITIVITIES

1. **Read labels carefully:** Allergens must be clearly labeled on packaged foods in many countries.
2. **Ask about ingredients when eating out:** Don't be shy - your health is important!
3. **Carry emergency medication:** If prescribed, always have your epinephrine auto-injector with you.
4. **Try elimination diets:** Under a doctor's supervision, remove suspected trigger foods and slowly reintroduce them to name sensitivities.

5. **Consider allergy testing:** Skin prick tests or blood tests can help name allergies.
6. **Stay informed:** Food manufacturing processes can change, so always check labels, even on familiar products.
7. **Prepare for travel:** Research local cuisines and learn how to communicate your allergies in the local language.
8. **Educate others:** Instruct friends and family about your allergies so they can help keep you safe.

EMERGING TREATMENTS

1. **Oral Immunotherapy (OIT):** Gradually introducing insignificant amounts of an allergen to build tolerance.
2. **Epicutaneous Immunotherapy (EPIT):** Using a skin patch to expose the immune system to tiny amounts of an allergen.
3. **Biologics:** Drugs that target specific parts of the allergic response, like the antibody Xolair for certain food allergies.

THE FUTURE OF FOOD ALLERGY RESEARCH

Scientists are working hard to understand food allergies better and develop new treatments. Some exciting areas of research include:

1. **Microbiome studies:** Looking at how gut bacteria might influence allergy development.
2. **Genetic research:** Naming genes that might make someone more likely to develop allergies.

3. **Early introduction:** Studying how introducing potential allergens early in life might prevent allergies from developing.
4. **Cross-reactivity:** Understanding why some people with certain allergies (like birch pollen) react to related foods (like apples).

Remember: Just because you're not allergic to a food doesn't mean you can't be sensitive to it. Listen to your body!

LIVING WITH FOOD ALLERGIES: A PERSONAL STORY

Meet Sarah, a 12-year-old with severe peanut and tree nut allergies. She was diagnosed as a toddler after a scary reaction to a peanut butter sandwich. Now, Sarah and her family have learned to navigate life with food allergies:

- They read every label, even on familiar products.
- Sarah wears a medical alert bracelet and carries two epinephrine auto-injectors.
- The family has taught Sarah's friends about her allergies and how to use her auto-injector in an emergency.
- They've found nut-free alternatives for Sarah's favorite foods, like sunflower seed butter instead of peanut butter.
- Sarah's school has a nut-free policy to help keep her and other allergic students safe.

Living with food allergies can be challenging, but with the right precautions, people like Sarah can lead full, active lives.

WRAPPING UP

Whew! We've covered a lot of ground in our journey through food allergies and sensitivities. From the science behind allergic reactions to the latest treatments, we've seen how complex and individual these issues can be. Remember, if you suspect you have a food allergy or sensitivity, it's important to talk to a doctor or allergist. They can help you get accurately diagnosed and develop a management plan.

Living with food allergies or sensitivities might seem overwhelming at first, but with knowledge and preparation, it's manageable. Who knows - maybe you'll discover some new favorite foods as you explore alternatives!

So, the next time you're at a potluck or trying a new restaurant, think about what you've learned. A little awareness goes a long way in keeping everyone safe and healthy. And hey, being considerate of others' dietary needs? That's simply good manners, and it might even save a life. Now go forth and eat wisely, food explorers!

CHAPTER FOUR
ENVIRONMENTAL CONTAMINANTS IN FOOD - UNINVITED GUESTS ON YOUR PLATE

All right, food sleuths, it's time to put on our detective hats and dive into the mysterious world of environmental contaminants in our food. These are the sneaky substances that find their way into our meals without an invitation. Don't worry, though - by the end of this chapter, you'll be a pro at spotting these culprits and knowing how to deal with them!

WHAT ARE ENVIRONMENTAL CONTAMINANTS?

Environmental contaminants are substances that get into our food from the environment around us. They can come from natural sources, industrial processes, or even from the way we grow and handle our food. Let's break them down into categories:

1. HEAVY METALS

Heavy metals are elements that can be toxic to our bodies in considerable amounts. Some common culprits include:

• MERCURY

- **Sources:** Mainly from industrial pollution that gets into water
- **Foods often affected:** Large predatory fish like tuna, swordfish, and shark
- **Health effects:** Can damage the nervous system, especially in developing fetuses

• LEAD

- **Sources:** Old paint, contaminated soil, some ceramics
- **Foods often affected:** Fruits and vegetables grown in contaminated soil, water from old pipes
- **Health effects:** Can cause developmental problems in children, kidney damage in adults

• CADMIUM

- **Sources:** Industrial processes, cigarette smoke, fertilizers
- **Foods often affected:** Shellfish, kidney meats, some plants (especially leafy greens and root vegetables)
- **Health effects:** Can damage kidneys and weaken bones

• ARSENIC

- **Sources:** Naturally occurring in soil and water, pesticides
- **Foods often affected:** Rice, some fruit juices, seafood

- **Health effects:** Increased cancer risk, skin lesions

MERCURY IN FISH: A CLOSER LOOK

Let's dive deeper into the issue of mercury in fish, as it's a common concern:

- **Bioaccumulation:** Mercury builds up in fish over time, so older, larger fish tend to have more.
- **Types of fish to limit:** Shark, swordfish, king mackerel, and tilefish
- **Lower-mercury options:** Salmon, cod, tilapia, and canned light tuna
- **Special precautions:** Pregnant women, nursing mothers, and young children should be extra careful about their fish consumption

Did You Know? The FDA and EPA provide guidelines for fish consumption based on mercury levels. They recommend eating 2-3 servings of low-mercury fish per week for best health benefits.

2. PERSISTENT ORGANIC POLLUTANTS (POPS)

POPs are nasty chemicals that stick around in the environment for a long time and can accumulate in the food chain. Some examples include:

PCBS (POLYCHLORINATED BIPHENYLS)
- Once used in electrical equipment

- Can still be found in some fish from contaminated waters
- Health effects: Possible carcinogen, can affect immune and reproductive systems

• DIOXINS

- Byproducts of industrial processes and burning
- Found mainly in animal fats
- **Health effects:** Can cause reproductive and developmental problems, damage the immune system, and cause cancer

• DDT (DICHLORODIPHENYLTRICHLOROETHANE)

- A pesticide banned in many countries but persists in the environment
- Can be found in fatty animal products
- **Health effects:** Possible carcinogen, may affect reproductive system

Fun Fact: Some plants, like pumpkins and zucchinis, are good at absorbing POPs from soil. Scientists are exploring using these plants to clean up contaminated areas - a process called phytoremediation!

3. MICROPLASTICS

These tiny pieces of plastic (less than 5mm (about 0.2 in) in size) are becoming a big problem in our environment and our food.

- **Sources:** Breakdown of larger plastic waste, microbeads from personal care products
- **Foods often affected:** Seafood, salt, bottled water
- **Health concerns:** Still being studied, but may include carrying other contaminants and affecting digestive system

Shocking Stat: A 2019 study by WWF International suggested that people might be ingesting about 5 grams of plastic every week - that's about the weight of a credit card!

4. PESTICIDE RESIDUES

Pesticides help protect crops from pests, but residues can remain on our food.

- **Types:** Insecticides, herbicides, fungicides
- **Foods often affected:** Fruits and vegetables, especially those with edible skins
- **Health concerns:** Some pesticides are linked to cancer, hormone disruption, and neurological issues

THE "DIRTY DOZEN" AND "CLEAN FIFTEEN"

Each year, the Environmental Working Group releases lists of produce with the most and least pesticide residues. Here's the 2021 list:

DIRTY DOZEN (HIGHEST IN PESTICIDES)

1. Strawberries
2. Spinach

3. Kale, collard and mustard greens
4. Nectarines
5. Apples
6. Grapes
7. Cherries
8. Peaches
9. Pears
10. Bell and hot peppers
11. Celery
12. Tomatoes

CLEAN FIFTEEN (LOWEST IN PESTICIDES)

1. Avocados
2. Sweet corn
3. Pineapple
4. Onions
5. Papaya
6. Sweet peas (frozen)
7. Eggplant
8. Asparagus
9. Broccoli
10. Cabbage
11. Kiwi
12. Cauliflower
13. Mushrooms
14. Honeydew melon
15. Cantaloupes

Real-world Study: A 2019 study published in the journal Environmental Health Perspectives found that switching to an organic diet for just one week reduced the levels of pesticides in participants' bodies by an average of 60%!

5. NATURAL TOXINS

Not all food contaminants come from human activities. Some plants produce their own natural toxins as a defense mechanism.

- **Solanine:** Found in green parts of potatoes and green tomatoes
- **Cyanogenic glycosides:** In apple seeds and some stone fruit pits
- **Aflatoxins:** Produced by certain molds on peanuts, corn, and other crops

Fun Fact: The bitter taste in some foods is often from natural toxins. That's why animals in the wild often avoid bitter plants!

HOW DO CONTAMINANTS GET INTO OUR FOOD?

Understanding how these uninvited guests crash our food Certainly! Let's continue with how contaminants get into our food and then move on to the impacts and ways to reduce exposure.

1. AIR POLLUTION

- Contaminants in the air can settle on crops or be absorbed by plants
- **Example:** Dioxins from industrial emissions can land on grazing land and end up in meat and dairy products

2. WATER CONTAMINATION

- Pollutants in water can be absorbed by fish or used to irrigate crops
- **Example:** Mercury from industrial runoff can accumulate in fish

3. SOIL CONTAMINATION

- Plants can absorb contaminants from the soil through their roots
- **Example:** Lead from old paint or industrial sites can end up in vegetables grown in contaminated soil

4. FOOD PROCESSING AND PACKAGING

- Some contaminants can be introduced during food processing or leach from packaging
- **Example:** BPA from linings can migrate into canned foods

5. BIOACCUMULATION

- Some contaminants build up in the food chain, becoming more concentrated in animals at the top

- **Example:** PCBs accumulate in the fatty tissues of fish, with larger predatory fish having higher levels

IMPACT ON HUMAN HEALTH

The health effects of environmental contaminants can vary widely depending on the specific contaminant, the level of exposure, and individual factors. Some potential impacts include:

1. **Cancer:** Some contaminants, like certain pesticides and heavy metals, are known or suspected carcinogens.
2. **Reproductive Issues:** Certain contaminants can affect fertility or fetal development.
3. **Neurological Problems:** Heavy metals like lead and mercury can damage the nervous system.
4. **Hormone Disruption:** Some chemicals, known as endocrine disruptors, can interfere with hormone function.
5. **Immune System Effects:** Certain contaminants can weaken the immune system, making people more susceptible to illness.

CASE STUDY: THE MINAMATA DISASTER

One of the most infamous cases of food contamination occurred in Minamata, Japan, in the 1950s. A chemical factory was dumping mercury-contaminated wastewater into the bay, which accumulated in fish and shellfish. Local people who ate seafood from the bay developed severe neurological symptoms, now known as Minamata disease. This disaster highlighted the dangers of industrial pollution and the importance of environmental protection.

REDUCING EXPOSURE TO CONTAMINANTS

While it's impossible to completely avoid all contaminants, there are steps we can take to reduce our exposure:

1. **Eat a Varied Diet:**
   - Eating a diverse range of foods helps prevent overexposure to any one contaminant
   - Rotate types of fish, grains, and produce to get a balance of nutrients and minimize risks

2. **Wash and Prepare Food Properly:**
   - Wash fruits and vegetables thoroughly, even if you plan to peel them
   - Peel fruits and vegetables, when possible, especially if they're not organic
   - Trim fat from meat and remove skin from fish, as some contaminants accumulate in fat

3. **Choose Organic When Possible:**
   - Organic produce generally has lower levels of synthetic pesticide residues
   - Focus on buying organic versions of the "Dirty Dozen" if budget is a concern

4. **Be Aware of Local Advisories:**
   - Pay attention to local fish advisories, especially for recreational fishing
   - Be cautious about eating locally grown produce if you live in an area with known soil contamination

5. **Use Safe Cooking and Storage Practices:**
   - Avoid using old or scratched non-stick cookware
   - Store food in glass or stainless-steel containers rather than plastic when possible
   - Don't microwave food in plastic containers

6. **Filter Your Water:**
   - Use a water filter certified to remove contaminants of concern in your area
   - If you have well water, have it tested regularly for contaminants

7. **Stay Informed:**
   - Keep up with the latest research and recommendations from reputable health organizations
   - Be critical of sensationalized reports about food dangers – always check reliable sources

THE ROLE OF REGULATION

Government agencies play a crucial role in monitoring and regulating contaminants in our food supply. In the United States, for example:

- The **FDA** (Food and Drug Administration) sets and enforces standards for contaminants in food.
- The **EPA** (Environmental Protection Agency) regulates pesticides and sets standards for drinking water.
- The **USDA** (United States Department of Agriculture) oversees organic certification and meat inspection.

These agencies conduct regular testing, set safety limits for contaminants, and can issue recalls when necessary. However, it's important to note that regulations can vary by country, and emerging contaminants may not yet be fully regulated.

GLOBAL EFFORTS TO REDUCE CONTAMINANTS

Addressing food contamination is a global issue. International organizations like the WHO (World Health Organization) and FAO (Food and Agriculture Organization) work to:

- Set international standards for food safety
- Coordinate research on contaminants
- Help countries develop better food safety systems

One important international agreement is the Stockholm Convention on Persistent Organic Pollutants, which aims to eliminate or restrict the production and use of POPs globally.

THE FUTURE OF FOOD SAFETY

Scientists and researchers are continually working to better understand and address food contamination. Some exciting developments include:

1. **Improved Detection Methods:**
   - More sensitive tests can detect contaminants at lower levels
   - Rapid testing methods allow for quicker identification of problems

2. **Sustainable Agriculture Practices:**
   - Methods like integrated pest management can reduce the need for pesticides
   - Crop rotation and other techniques can help keep soil health

3. **Bioremediation:**
   - Using plants or microorganisms to clean up contaminated soil and water

4. **Nanotechnology:**
   - Developing packaging materials that can detect and even neutralize contaminants

5. **Gene Editing:**

- Creating crop varieties that are naturally resistant to pests and diseases, reducing the need for pesticides

We've covered a lot of ground in our exploration of environmental contaminants in food! It might seem a bit overwhelming, but remember:

- Our food supply is generally very safe, especially in countries with strong regulations.
- The benefits of eating a varied diet rich in fruits and vegetables far outweigh the risks of contaminant exposure for most people.
- Minor changes in how we choose, prepare, and store food can significantly reduce our exposure to contaminants.

By staying informed and making smart choices, we can enjoy the benefits of a healthy diet while minimizing our exposure to environmental contaminants. So, the next time you're at the grocery store or farmers market, think about what you've learned. You're now equipped with the knowledge to make informed decisions about the food you eat.

Remember, being aware doesn't mean being afraid. It means being empowered to make the best choices for your health and the environment. Now go out there and enjoy your food, knowing you're making smart, informed decisions!

CHAPTER FIVE
FOOD PACKAGING AND CHEMICAL MIGRATION

WHEN THE CONTAINER BECOMES THE CONTENT

Hey there, food detectives! Ready to unwrap the mystery of food packaging? You might think the wrapper is just there to keep your snacks fresh, but there's a whole lot more going on. Let's dive into the world of food packaging and learn about how sometimes, what's on the outside can get on the inside of our food!

WHAT IS FOOD PACKAGING?

Food packaging is like a superhero suit for your food. It protects food from damage, keeps it fresh, and gives us valuable information about what we're eating. But just like how superheroes can sometimes cause accidental damage while saving the day, food packaging can sometimes add things to our food that we didn't expect.

COMMON FOOD PACKAGING MATERIALS

1. PLASTIC

- **Types:** PET, HDPE, PVC, LDPE, PP, PS
- **Used for:** Bottles, bags, containers, wraps
- **Pros:** Lightweight, cheap, versatile
- **Cons:** Some can leach chemicals, environmental concerns

2. METAL

- **Types:** Aluminum, steel
- **Used for:** Cans, foil
- **Pros:** Durable, good barrier properties
- **Cons:** Can leach tiny amounts of metals, especially with acidic foods

3. GLASS

- **Used for:** Bottles, jars
- **Pros:** Inert (doesn't react with food), recyclable
- **Cons:** Heavy, breakable

4. PAPER AND CARDBOARD

- **Used for:** Boxes, cartons, paper plates
- **Pros:** Biodegradable, easy to print on
- **Cons:** Not moisture-resistant unless coated

5. STYROFOAM (EXPANDED POLYSTYRENE)

- **Used for:** Takeout containers, cups

- **Pros:** Good insulator, lightweight
- **Cons:** Environmental concerns, can leach styrene

Let's dive deeper into each of these:

1. PLASTIC: THE SHAPESHIFTER OF PACKAGING

Plastics are everywhere in food packaging, and for good reason - they're cheap, lightweight, and can be molded into any shape. But not all plastics are created equal. Here's a rundown of common types:

• PET (POLYETHYLENE TEREPHTHALATE)

- **Used for:** Water bottles, soda bottles
- **Recycling code:** 1
- **Safety:** Generally considered safe, but can leach antimony (a toxic metal) in tiny amounts, especially when heated

• HDPE (HIGH-DENSITY POLYETHYLENE)

- **Used for:** Milk jugs, juice bottles
- **Recycling code:** 2
- **Safety:** Considered one of the safest plastics

• PVC (POLYVINYL CHLORIDE)

- **Used for:** Some cling wraps
- **Recycling code:** 3
- **Safety:** Can leach phthalates and lead, best avoided for food

- ## LDPE (LOW-DENSITY POLYETHYLENE)
  - **Used for:** Squeezable bottles, some plastic bags
  - **Recycling code:** 4
  - **Safety:** Considered safe, doesn't leach chemicals

- ## PP (POLYPROPYLENE)
  - **Used for:** Yogurt containers, bottle caps
  - **Recycling code:** 5
  - **Safety:** Considered safe, heat-resistant

- ## PS (POLYSTYRENE)
  - **Used for:** Disposable plates, cups
  - **Recycling code:** 6
  - **Safety:** Can leach styrene, especially when heated

Fun Fact: The recycling codes on plastic items (those numbers inside the triangle of arrows) aren't just for recycling - they also tell you what type of plastic it is!

2. METAL: THE TOUGH GUY OF PACKAGING

Metal packaging has been around for over 200 years and is still going strong. Let's look at the two main types:

- ## ALUMINUM
  - **Used for:** Soda cans, foil
  - **Pros:** Lightweight, good barrier to light and oxygen
  - **Cons:** Can leach lesser amounts of aluminum, especially with acidic foods

- **STEEL**
  - **Used for:** Food cans, aerosol cans
  - **Pros:** Strong, recyclable
  - **Cons:** Can rust if the protective coating is damaged

Did You Know? The inside of most metal cans is coated with a thin layer of plastic or resin to prevent the metal from contacting the food directly.

3. GLASS: THE OLD RELIABLE

Glass has been used for food storage for thousands of years, and for good reason:

- **PROS**
  - Doesn't react with food
  - Can be heated without leaching chemicals
  - Infinitely recyclable without loss of quality

- **CONS**
  - Heavy
  - Breakable
  - Energy-intensive to produce

Cool Chemistry: Glass is made from melting sand (silica) with other minerals. Its molecular structure is what makes it transparent and nonreactive.

4. PAPER AND CARDBOARD: THE NATURAL CHOICE?

Paper-based packaging seems eco-friendly, but it's not always as simple as it looks:

• TYPES

- Paperboard (cereal boxes)
- Corrugated cardboard (shipping boxes)
- Paper bags

• PROS

- Biodegradable
- Made from renewable resources
- Easy to recycle

• CONS

- Not waterproof unless coated
- Coatings can include plastic or other chemicals
- Can absorb printing inks, which might contain harmful substances

Eco Fact: While paper is biodegradable, it can release methane (a potent greenhouse gas) when it breaks down in landfills without oxygen.

5. STYROFOAM: THE CONTROVERSIAL INSULATOR

Styrofoam, or expanded polystyrene, is great at keeping hot things hot and cold things cold, but it comes with some drawbacks:

• PROS

- Excellent insulator
- Lightweight
- Cheap to produce

• CONS

- Difficult to recycle
- Can leach styrene, especially when heated
- Persists in the environment for a long time

Environmental Impact: Scientists estimate that Styrofoam can take up to 500 years to decompose in the environment!

CHEMICAL MIGRATION: THE SNEAKY MOVE

Now that we know about different packaging materials, let's talk about how they can affect our food. Chemical migration is when substances from the packaging move into the food. It's like when you leave a slice of pizza in a plastic container and the container smells like pizza afterwards - but in reverse, and on a molecular level!

Factors that Affect Chemical Migration:

1. **Heat:** Higher temperatures speed up chemical reactions and can cause more migration.
2. **Time:** The longer food is in contact with packaging, the more migration can occur.
3. **Fat Content:** Fatty foods tend to absorb more chemicals from plastic.
4. **Acidity:** Acidic foods can cause more leaching from some packaging materials.
5. **Physical Contact:** More contact between food and packaging means more potential for migration.

Let's look at some chemicals of concern:

1. BISPHENOL A (BPA)

- **Used in:** Some plastic containers and can linings
- **Concerns:** Can act like estrogen in the body
- **Health effects:** Linked to reproductive issues, certain cancers, and developmental problems
- **Alternatives:** Many companies now use "BPA-free" alternatives, but some of these (like BPS) may have similar effects

BPA Study: A 2018 study published in the journal Hypertension found that BPA exposure was associated with an increased risk of high blood pressure in children and adolescents.

2. PHTHALATES

- **Used in:** Some food packaging plastics to make them flexible
- **Concerns:** Can disrupt hormones
- **Health effects:** Reproductive issues, developmental problems
- **Common sources:** Plastic wrap, plastic bags, some beverage containers

3. PER- AND POLYFLUOROALKYL SUBSTANCES (PFAS)

- **Used in:** Grease-resistant food packaging (like microwave popcorn bags and fast-food wrappers)
- **Concerns:** Very persistent in the environment and the human body
- **Health effects:** Linked to cancer, thyroid disease, immune system problems
- **Nickname:** "Forever chemicals" because they don't break down easily in the environment

PFAS Fact: A 2020 study found PFAS in the breast milk of US women at levels nearly 2,000 times higher than what's considered safe in drinking water.

4. STYRENE

- **Used in:** Styrofoam containers
- **Concerns:** Can leach into food, especially when heated
- **Health effects:** Possible human carcinogen, can affect the nervous system

- **Migration:** More likely to occur with hot, fatty foods

5. PRINTING INKS AND ADHESIVES

- **Used on:** Paper and cardboard packaging
- **Concerns:** Some inks and adhesives have harmful chemicals
- **Health effects:** Vary depending on the specific chemicals, but can include hormone disruption and cancer risk
- **Lesser-known risk:** These can be an issue even when not in direct contact with food, as chemicals can migrate through packaging layers

Mineral Oil Study: A 2017 study found that mineral oils from printing inks on cardboard food packaging could migrate into food, even though protective inner bags.

6. ANTIMONY

- **Found in:** PET plastic bottles
- **Concerns:** Can leach in insignificant amounts, especially in warm conditions
- **Health effects:** In high doses, can cause nausea, vomiting, and diarrhea
- **Migration:** Increases with higher temperatures and longer storage times

Real-world Example: A 2018 study published in the journal Water Research found that leaving PET bottled water in a car on a sweltering day significantly increased the levels of antimony in the water.

REDUCING CHEMICAL MIGRATION

1. Avoid microwaving food in plastic containers
2. Store food in glass or stainless steel when possible
3. Don't reuse single-use plastic bottles
4. Let hot food cool before putting it in plastic containers
5. Choose safer plastics: Look for recycling codes 1, 2, 4, and 5
6. Limit use of canned foods, especially for acidic items like tomatoes
7. Remove food from cans after opening and store in glass or ceramic containers

INNOVATIONS IN FOOD PACKAGING

Scientists and companies are always working on new ways to make food packaging safer and more sustainable. Here are some cool innovations:

1. EDIBLE PACKAGING

- Made from natural materials like seaweed or milk proteins
- Can be eaten along with the food or will biodegrade quickly
- **Example:** Ooho! water pouches made from seaweed extract

2. ACTIVE PACKAGING

- Interacts with food to extend shelf life
- Can absorb oxygen or release preservatives
- **Example:** Packets that absorb ethylene gas to keep fruits fresh longer

3. INTELLIGENT PACKAGING

- Has sensors to check food quality
- Can change color to show freshness
- **Example:** Labels that change color when milk spoils

4. BIOPLASTICS

- Made from renewable resources like corn starch or sugarcane
- Biodegradable and compostable
- **Example:** PLA (polylactic acid) used for some disposable cups and utensils

5. NANOCOMPOSITES

- Packaging materials with nanoparticles to improve properties
- Can provide better barriers against oxygen and moisture
- **Example:** Nanocomposite films that extend the shelf life of cheese

ECO-FRIENDLY PACKAGING: THE GREEN REVOLUTION

As we become more aware of environmental issues, many companies are trying to make their packaging eco-friendlier. Here are some approaches:

1. REDUCED PACKAGING

- Using less material overall
- **Example:** Concentrated laundry detergents in smaller bottles

2. RECYCLABLE PACKAGING

- Designing packaging to be easily recycled
- **Example:** Mono-material pouches that don't mix diverse types of plastic

3. COMPOSTABLE PACKAGING

- Breaks down into natural materials in composting conditions
- **Example:** Compostable chip bags made from plant-based materials

4. REUSABLE PACKAGING

- Designed to be used multiple times
- **Example:** Loop's system of reusable containers for common household products

5. PLASTIC-FREE PACKAGING

- Using alternatives to plastic
- **Example:** Boxed water in paper cartons instead of plastic bottles

THE ROLE OF CONSUMERS

As consumers, we play a crucial role in driving changes in food packaging. Here's how you can make a difference:

1. Choose products with minimal or eco-friendly packaging
2. Properly recycle or dispose of packaging
3. Use reusable containers and bags when shopping
4. Support companies that prioritize safe and sustainable packaging
5. Stay informed about packaging issues and share knowledge with others

REGULATIONS AND SAFETY STANDARDS

Government agencies set rules to keep our food packaging safe. In the U.S., the FDA (Food and Drug Administration) regulates food contact materials. They must approve new packaging materials before they can be used with food.

In the European Union, the European Food Safety Authority (EFSA) plays a similar role. They often have stricter standards than the U.S. For example, they've banned some chemicals that are still allowed in U.S. food packaging.

Global Issue: Different countries have different rules about food packaging. This can be challenging for companies that sell products internationally.

THE FUTURE OF FOOD PACKAGING

As we learn more about chemical migration and environmental impacts, food packaging will keep evolving. Here are some trends to watch:

1. More plant-based materials
2. Advanced recycling technologies
3. Packaging that communicates digitally (like QR codes with product info)
4. Increased use of nanotechnology for better food preservation
5. Stricter regulations on chemicals in food packaging

Wow, we've really unboxed the world of food packaging! From the materials used to the chemicals that might be hitching a ride into our food, there's a lot to consider when we're choosing our groceries.

Remember, while it's good to be aware of these issues, don't let it stress you out too much. Our food supply is generally very safe, and regulators and scientists are always working to make it safer. By making informed choices and handling food and packaging properly, we can minimize our exposure to potentially harmful chemicals.

So, the next time you're unwrapping a snack or storing leftovers, think about what you've learned. You're now equipped with the knowledge to make smart decisions about food packaging. And who knows? Maybe you'll be the one to invent the next important thing in safe, sustainable food packaging!

PART V
SPECIAL TOPICS IN FOOD CHEMISTRY

Welcome back, food chemistry explorers! We're about to dive even deeper into the fascinating worlds of organic food, GMOs, and the chemistry of cooking. Get ready for a feast of knowledge that will change the way you think about your food!

CHAPTER SIX
ORGANIC FOOD: MYTHS AND REALITIES

THE ORGANIC REVOLUTION

Organic food has gone from a niche market to a mainstream phenomenon. In 2020, organic food sales in the U.S. reached a record $56.4 billion (about $170 per person in the US) (about $170 per person in the US). But what's behind this growth, and does organic live up to the hype?

WHAT MAKES FOOD ORGANIC?

Let's break down the official requirements in more detail:

1. SOIL MANAGEMENT

- Must use natural fertilizers like compost or manure
- No synthetic fertilizers allowed
- Crop rotation to manage soil nutrients

2. PEST CONTROL

- Focus on prevention through habitat management
- Can use natural pesticides as a last resort

- No synthetic pesticides allowed

3. WEED CONTROL

- Mechanical or hand weeding
- Mulching to suppress weeds
- No synthetic herbicides allowed

4. LIVESTOCK MANAGEMENT

- Animals must have access to outdoors
- Organic feed only
- No antibiotics or growth hormones

5. PROCESSING

- Minimal processing preferred
- No artificial colors, flavors, or preservatives
- No irradiation or genetic engineering

ORGANIC CERTIFICATION PROCESS

1. Farm develops an organic system plan
2. Farm implements organic practices for 3 years
3. USDA-accredited certifier inspects the farm
4. Annual inspections to keep certification

ORGANIC MYTH #1: ORGANIC FOOD IS PESTICIDE-FREE

Reality Check: While organic farms use fewer pesticides, they're not pesticide-free. Let's look closer at organic pest control:

NATURAL PESTICIDES ALLOWED IN ORGANIC FARMING

- **Neem oil:** Derived from neem tree seeds, effective against many insects
- **Pyrethrin:** From chrysanthemum flowers, broad-spectrum insecticide
- **Copper sulfate:** Used to control fungal and bacterial diseases
- **Spinosad:** Produced by soil bacteria, controls caterpillars and flies
- **Diatomaceous earth:** Fossilized algae, damages insects' exoskeletons

While these are generally considered safer than synthetic pesticides, they're not without risks. For example, copper sulfate can accumulate in soil and harm beneficial soil organisms.

INTEGRATED PEST MANAGEMENT (IPM) IN ORGANIC FARMING

1. **Prevention:** Crop rotation, companion planting
2. **Monitoring:** Regular checks for pest problems
3. **Mechanical controls:** Traps, barriers

4. **Biological controls:** Introducing beneficial insects
5. **Chemical controls:** Natural pesticides as a last resort

ORGANIC MYTH #2: ORGANIC FOOD IS MORE NUTRITIOUS

Reality Check: The nutritional differences between organic and conventional foods are generally small. Let's break it down:

POTENTIAL NUTRITIONAL ADVANTAGES OF ORGANIC

- Higher levels of certain antioxidants (20-40% higher in some studies)
- Slightly higher omega-3 fatty acids in organic milk and meat
- Lower cadmium levels in some organic grains

POTENTIAL NUTRITIONAL DISADVANTAGES OF ORGANIC

- Sometimes lower protein content in organic grains
- Can have lower selenium levels (depends on soil content)

WHY THE DIFFERENCES?

- **Stress Response:** Without synthetic pesticides, plants may produce more antioxidants as a natural defense.
- **Fertilizer Use:** Conventional nitrogen fertilizers can lead to faster growth but lower nutrient density.
- **Soil Health:** Organic practices may lead to healthier soil, potentially affecting nutrient uptake.

META-ANALYSIS SPOTLIGHT

A 2014 study in the British Journal of Nutrition analyzed 343 peer-reviewed publications. They found that organic crops had:

- 18-69% higher concentrations of antioxidants
- Lower concentrations of cadmium (-48%)
- Lower nitrogen content
- Lower residues of pesticides

However, the authors noted that the differences, while statistically significant, were relatively small in terms of overall diet.

ORGANIC MYTH #3: ORGANIC FARMING IS BETTER FOR THE ENVIRONMENT

Reality Check: Organic farming has both environmental benefits and challenges. Let's dive deeper:

ENVIRONMENTAL BENEFITS OF ORGANIC FARMING

1. **Soil Health:**
    - Builds organic matter in soil
    - Promotes beneficial soil microorganisms
    - Reduces soil erosion

2. **Water Quality:**
    - Reduces nitrogen runoff
    - Cuts synthetic pesticide contamination

3. **Biodiversity:**
   - Supports more diverse ecosystems
   - Benefits pollinators and other beneficial insects

4. **Climate Change:**
   - Organic soils can sequester more carbon
   - Lower energy inputs (no synthetic fertilizer production)

ENVIRONMENTAL CHALLENGES OF ORGANIC FARMING

1. **Land Use:**
   - Generally lower yields mean more land needed for same output
   - Could lead to deforestation if scaled up globally

2. **Greenhouse Gas Emissions:**
   - Some practices, like tillage for weed control, can release soil carbon
   - Organic cattle might produce more methane due to diet

3. **Nutrient Management:**
   - Risk of nutrient runoff from manure
   - Potential for nutrient deficiencies in soil over time

4. **Pest Control:**
   - Some natural pesticides can harm beneficial insects
   - May require more applications, increasing fuel use

CASE STUDY: ORGANIC VS. CONVENTIONAL APPLE PRODUCTION

A life cycle assessment published in the Journal of Cleaner Production compared organic and conventional apple production in Washington State. They found:

- Organic apples had lower energy use and global warming potential per kilogram of fruit.
- Conventional apples had lower land use and water use per kilogram.
- Organic apples had higher potential for toxicity to aquatic ecosystems due to use of copper-based fungicides.

This study highlights the complexity of comparing environmental impacts of organic and conventional farming.

THE BOTTOM LINE ON ORGANIC: AN EXPANDED PERSPECTIVE

Choosing organic is a nuanced decision that depends on individual priorities:

HEALTH CONSIDERATIONS

- Reduces exposure to synthetic pesticides
- May have slightly higher levels of certain nutrients
- Not necessarily "healthier" overall

ENVIRONMENTAL CONSIDERATIONS

- Generally, better for soil health and biodiversity
- Can have lower carbon footprint, but not always
- Challenges with land use efficiency

ECONOMIC CONSIDERATIONS

- Often more expensive for consumers
- Can provide better income for farmers
- Supports a different model of agriculture

REAL-LIFE STORY: THE ORGANIC CONVERT

Let's revisit Sarah, the mom who switched to organic food when her daughter was born. Her journey didn't stop at just buying organic:

"After learning more about organic farming, I started volunteering at a local organic farm," Sarah says. "I saw firsthand how they managed pests without synthetic pesticides. It was eye-opening to see the amount of work that goes into it."

Sarah's experience changed her perspective: "I realized organic wasn't just about what's not in the food, but about supporting a whole system of agriculture. Now, I prioritize local and seasonal foods, whether they're certified organic or not. I've even started my own vegetable garden!"

Sarah's story illustrates how understanding organic food can lead to broader changes in how we think about and interact with our food system.

CHAPTER SEVEN
GENETICALLY MODIFIED ORGANISMS (GMOS) AND THEIR CHEMICAL PROFILES

Genetically modified organisms have been part of our food system since the 1990s, but they still are a topic of intense debate. Let's dive deeper into the science, controversy, and future of GMOs.

GMO 101: A CLOSER LOOK AT THE SCIENCE

Genetic modification involves altering an organism's DNA in a way that doesn't occur through traditional breeding. Here are the main techniques:

1. TRANSGENESIS

- Inserting genes from one species into another
- **Example:** Bt corn, which contains a gene from bacteria

2. CIS GENESIS

- Inserting genes from the same or closely related species
- **Example:** Disease-resistant apples using genes from wild apples

3. GENE SILENCING

- Turning off specific genes
- **Example:** Non-browning Arctic Apples

4. GENE EDITING (NEWER TECHNIQUE)

- Making precise changes to existing DNA
- **Example:** CRISPR-edited mushrooms that don't brown

COMMON GMO CROPS: A DEEPER DIVE

1. CORN (BT AND HERBICIDE-RESISTANT)

- Bt corn produces its own insecticide
- Herbicide-resistant corn tolerates glyphosate
- 92% of U.S. corn was genetically modified in 2020

2. SOYBEANS (HERBICIDE-RESISTANT)

- Tolerates glyphosate herbicide
- 94% of U.S. soybeans were GM in 2020

3. COTTON (BT AND HERBICIDE-RESISTANT)

- Produces its own insecticide and tolerates herbicides
- 94% of U.S. cotton was GM in 2020

4. CANOLA (HERBICIDE-RESISTANT)

- Tolerates specific herbicides
- About 95% of Canadian canola is GM

5. PAPAYA (VIRUS-RESISTANT):

- Resistant to ringspot virus
- Saved the Hawaiian papaya industry

6. POTATOES (NON-BROWNING, BLIGHT-RESISTANT)

- Reduced black spot bruising and asparagine content
- Resistant to late blight disease

HOW GMOS CHANGE FOOD CHEMISTRY

GMOs can alter the chemical profile of foods in several ways:

1. NEW PROTEINS

- Bt corn produces Cry proteins toxic to certain insects
- These proteins are broken down in human digestive system

2. CHANGED NUTRIENT LEVELS

- Golden Rice is engineered to produce beta-carotene
- High-oleic acid soybeans have a different fatty acid profile

3. ALTERED PLANT DEFENSES

- Some GM crops produce lower levels of natural toxins
- Example: Bt corn can have lower levels of Fumo nisin, a fungal toxin

4. HERBICIDE RESIDUES

- Herbicide-resistant crops may have higher herbicide residues
- Glyphosate residues are higher in some GM crops, but still within safety limits

GMO MYTH #1: GMOS ARE UNSAFE TO EAT

Reality Check: Extensive studies have found no evidence that approved GMOs are harmful to human health. Let's look at the evidence:

MAJOR SCIENTIFIC CONSENSUS

- World Health Organization
- American Medical Association
- National Academy of Sciences
- European Commission

All conclude that GM foods currently on the market are safe to eat.

KEY STUDIES

1. **2016 National Academies Report**
    - Reviewed over 900 studies and 20 years of data
    - Found no substantiated evidence of health risks from GMOs

2. **2018 EU Study**
    - Analyzed 20 years of GMO research

o   Concluded no scientific evidence of health risks

3. **Long-term Animal Studies**
   o   2-year study on GM corn in rats showed no negative health effects
   o   Multi-generational studies in animals show no reproductive or developmental issues

ADDRESSING COMMON CONCERNS

- **Allergies:** GM foods are tested for new allergens before approval
- **Cancer:** No credible evidence linking GM foods to cancer
- **Antibiotic Resistance:** Antibiotic resistance marker genes in GM crops haven't been shown to transfer to bacteria in gut

GMO MYTH #2: GMOS USE MORE PESTICIDES

Reality Check: The impact of GMOs on pesticide use is complex and varies by crop and region. Let's break it down:

INSECTICIDE USE

- Bt crops have led to significant reductions in insecticide use
- **Example:** Bt cotton in India reduced insecticide use by 41%

HERBICIDE USE

- Herbicide-resistant crops have led to increased use of specific herbicides, particularly glyphosate
- However, glyphosate has replaced other, more toxic herbicides

OVERALL TRENDS

- A 2014 meta-analysis found that GM technology has reduced pesticide use by 37%
- But the rise of herbicide-resistant weeds is leading to increased herbicide use in some areas

CASE STUDY: BT EGGPLANT IN BANGLADESH

- Introduced in 2014
- Reduced insecticide use by 61%
- Increased farmer income and reduced pesticide poisonings

GMO MYTH #3: GMOS ARE BAD FOR THE ENVIRONMENT

Reality Check: GMOs have both positive and negative environmental impacts. Let's explore further:

POSITIVE ENVIRONMENTAL IMPACTS

1. **Reduced Insecticide Use**
    - Bt crops have significantly reduced broad-spectrum insecticide use

- Helps non-target insects and promotes biodiversity

2. **Conservation Tillage**
    - Herbicide-resistant crops enable no-till farming
    - Reduces soil erosion and carbon emissions

3. **Land Use Efficiency**
    - Higher yields can reduce pressure to convert natural habitats to farmland
    - **Example:** GM cotton in India increased yields by 31%

4. **Reduced Food Waste**
    - Non-browning apples and potatoes could reduce food waste
    - Longer-lasting GM tomatoes reduce post-harvest losses

NEGATIVE ENVIRONMENTAL IMPACTS

1. **Herbicide Resistance**
    - Overuse of glyphosate has led to resistant weeds
    - Could lead to increased herbicide use over time

2. **Gene Flow**
    - Potential for GM traits to spread to non-GM crops or wild relatives
    - **Example:** GM canola found growing wild in North Dakota

3. **Impact on Non-Target Organisms**

- Bt toxins can affect some non-pest insects, though less than broad-spectrum insecticides
- Monarch butterfly populations have declined, partially due to loss of milkweed in GM crop fields

4. **Biodiversity Concerns**
   - Monoculture farming, often associated with GM crops, can reduce agricultural biodiversity

THE GMO DEBATE: A REAL-WORLD EXAMPLE

Let's revisit the case of GMO papaya in Hawaii, with more details:

1990s: Papaya Ringspot Virus threatens to wipe out Hawaii's papaya industry

1998: GM virus-resistant papaya approved and released

2000s: GM papaya saves the industry, accounting for 77% of Hawaii's papaya crop

But the story doesn't end there:

- **Export Challenges:** Japan, a major market, initially banned GM papaya
- **2011:** Japan approves import of GM papaya after years of testing
- **Organic Concerns:** GM pollen can contaminate non-GM and organic papaya farms
- **Cultural Issues:** Some Native Hawaiians oppose GMOs on cultural grounds

This example illustrates the complex interplay of science, economics, culture, and international trade in the GMO debate.

THE FUTURE OF GMOS: GENE EDITING AND BEYOND

Gene editing, especially CRISPR technology, is revolutionizing genetic modification:

POTENTIAL APPLICATIONS

- Gluten-free wheat
- Peanuts without allergens
- Disease-resistant cacao to save chocolate production
- High-fiber white rice

REGULATORY CHALLENGES

- U.S. considers most gene-edited crops as non-GM
- EU regulates them as GMOs
- Global regulatory disagreement could affect trade

EMERGING TECHNOLOGIES

- RNA interference to control pests without altering crop DNA
- Synthetic biology to produce food ingredients in labs

ETHICAL CONSIDERATIONS

- How to ensure fair access to GM technology

- Balancing innovation with precaution
- Transparency and consumer choice

CHAPTER EIGHT

THE IMPACT OF COOKING METHODS ON FOOD CHEMISTRY

THE MAGIC OF COOKING: A CHEMICAL TRANSFORMATION

Cooking is chemistry in action! Let's explore how different cooking methods change our food at a molecular level, and how we can use this knowledge to make healthier, tastier meals.

COOKING METHOD #1: BOILING

WHAT HAPPENS

- Water-soluble vitamins leach into cooking water
- Starches gelatinize, softening foods
- Proteins denature, changing texture

THE SCIENCE OF BOILING

- At sea level, water boils at 100°C (212°F)
- Heat causes starch granules to absorb water and burst
- Proteins unfold and reconnect in new shapes

NUTRIENT CHANGES

- Vitamin C losses can be up to 50%
- B vitamins can decrease by 30-40%
- Some minerals leach into water, but less than vitamins

BOILING TIPS

- Use minimal water to reduce nutrient loss
- Save nutrient-rich cooking water for soups or sauces
- Don't overcook to preserve nutrients and texture

Cool Chemistry (Pasta Science): When you salt pasta water, it raises the boiling point slightly. More importantly, it seasons the pasta from the inside out as it absorbs water during cooking.

COOKING METHOD #2: FRYING

WHAT HAPPENS

- Maillard reaction creates complex flavors
- Foods lose water and absorb fat
- Some nutrients are lost, others become more available

THE SCIENCE OF FRYING

- Oil temperatures typically range from 350-375°F (175-190°C)
- At these temperatures, water in food rapidly turns to steam, creating the characteristic sizzle
- The Maillard reaction occurs between amino acids and sugars, creating hundreds of flavor compounds

NUTRIENT CHANGES

- Fat-soluble vitamins (A, D, E, K) are kept and can become more bioavailable
- Water-soluble vitamins may be destroyed by heat
- Frying can increase the calorie content of food significantly

HEALTH CONSIDERATIONS

- Repeated heating of oil can create harmful compounds like acrylamide
- Trans fats can form when vegetable oils are heated to extremely elevated temperatures

FRYING TIPS

- Use oils with high smoke points like peanut or avocado oil for high-heat frying
- Don't reuse oil too many times
- Pat foods dry before frying to reduce oil splatter and absorption

Did You Know? The perfect french fry undergoes two frying stages: first at a lower temperature to cook the inside, then at a higher temperature to crisp the outside.

COOKING METHOD #3: GRILLING

WHAT HAPPENS

- Maillard reaction and caramelization create distinctive flavors

- Fat drips away, potentially making food less caloric
- Can create potentially harmful compounds like HCAs and PAHs

THE SCIENCE OF GRILLING

- Grilling typically occurs at temperatures between 300-500°F (150-260°C)
- Direct heat causes rapid browning on the surface
- Smoke from dripping fats adds flavor compounds

NUTRIENT CHANGES

- Can lead to significant vitamin loss, especially B vitamins
- Minerals are generally retained
- Grilling can increase the antioxidant content of some vegetables

HEALTH CONSIDERATIONS

- HCAs form when amino acids and creatine react at elevated temperatures
- PAHs form when fat drips onto hot coals and creates smoke
- Both HCAs and PAHs are potential carcinogens

GRILLING TIPS

- Marinate meat before grilling this can reduce HCA formation by up to 92%
- Avoid char by keeping food away from direct flames

- Grill more vegetables they don't produce HCAs when grilled

Grilling Study Spotlight: A study published in the Journal of Agricultural and Food Chemistry found that briefly microwaving meat before grilling reduced HCA formation by up to 95%.

COOKING METHOD #4: MICROWAVING

WHAT HAPPENS

- Foods cook quickly as water molecules are energized
- Can preserve more nutrients than other cooking methods
- Doesn't create Maillard reaction products

THE SCIENCE OF MICROWAVING

- Microwaves cause water molecules to vibrate, generating heat
- Cooking occurs from the outside in, but also throughout the food simultaneously
- No direct heat means no browning reactions

NUTRIENT CHANGES

- Generally, preserves nutrients better than other cooking methods due to shorter cooking times and less water used
- Can keep up to 90% of vitamin C in vegetables
- Doesn't significantly affect mineral content

MICROWAVE MYTHS DEBUNKED

- Microwaves don't make food radioactive
- They don't destroy nutrients any more than other cooking methods
- Microwave-safe plastics don't release harmful chemicals into food when used properly

MICROWAVING TIPS

- Use microwave-safe containers
- Stir food midway through cooking for even heating
- Let food stand for a minute after microwaving to allow heat to distribute evenly

Did You Know? The invention of the microwave oven was an accident. In 1945, engineer Percy Spencer noticed that a chocolate bar in his pocket melted while he was working with a magnetron, leading to the development of microwave cooking.

COOKING METHOD #5: PRESSURE COOKING
WHAT HAPPENS

- Higher pressure raises the boiling point of water, cooking food faster
- Can make foods more digestible by breaking down tough fibers
- May preserve more nutrients than boiling due to shorter cooking times

THE SCIENCE OF PRESSURE COOKING

- In a sealed pressure cooker, steam builds up, increasing pressure
- At 15 PSI (typical for most pressure cookers), water boils at 250°F (121°C)
- Higher temperature and pressure accelerate cooking processes

NUTRIENT CHANGES

- Can keep up to 90-95% of vitamins and minerals
- More effective at preserving heat-sensitive nutrients than regular boiling
- Can increase the bioavailability of some nutrients

PRESSURE COOKING BENEFITS

- Energy efficient due to shorter cooking times
- Can make tough cuts of meat more tender
- Allows cooking of beans and grains without pre-soaking

PRESSURE COOKING TIPS

- Don't overfill the cooker (usually no more than 2/3 full)
- Use enough liquid to create steam (usually at least 1 cup)
- Release pressure carefully to avoid burns

Cool Chemistry: At high altitudes, where water boils at lower temperatures, pressure cookers are especially useful for achieving proper cooking temperatures.

THE RAW FOOD DEBATE

Some people advocate for eating mostly raw foods, claiming it's more nutritious. Let's dive deeper into the science behind this debate:

PROS OF RAW FOOD:

- Preserves heat-sensitive vitamins like vitamin C and some B vitamins
- Retains food enzymes (though most are denatured by stomach acid anyway)
- Often lower in calories and higher in fiber
- May preserve certain phytochemicals that are altered by cooking

CONS OF RAW FOOD:

- Some nutrients become more available when cooked (like lycopene in tomatoes)
- Cooking kills potentially harmful bacteria and parasites
- Some foods are difficult to digest when raw (like starchy vegetables)
- Certain raw foods contain anti-nutrients that are reduced by cooking

RAW FOOD MYTHS AND FACTS:

- **Myth: Enzymes in raw food aid digestion**

Fact: Most food enzymes are denatured by stomach acid

- **Myth: Cooking destroys all nutrients**

Fact: While some nutrients are lost, others become more bioavailable

- **Myth: Raw diets cure diseases**

Fact: While a diet high in fruits and vegetables is healthy, there's no scientific evidence that raw diets cure diseases

RAW FOOD STUDY SPOTLIGHT

A study published in the British Journal of Nutrition found that following a strict raw food diet was associated with low bone mass and density, likely due to low calcium intake and vitamin D deficiency.

The bottom line: A mix of raw and cooked foods is probably best for most people, offering a balance of preserved nutrients and enhanced digestibility.

CASE STUDY: THE TOMATO TRANSFORMATION

Let's take a deeper look at how cooking changes the humble tomato:

RAW TOMATO

- High in vitamin C (about 14mg (about half the weight of a grain of rice) per 100g)
- Contains lycopene, but in a form, that's not easily absorbed
- Has a refreshing, slightly acidic taste
- Firm texture due to intact cell walls

COOKED TOMATO

- Loses some vitamin C (about 30% loss after 30 minutes of cooking)
- Lycopene becomes more bioavailable (up to 164% increase)
- Develops deeper, sweeter flavor due to concentration of sugars and acids
- Softer texture as heat breaks down cell walls

THE LYCOPENE FACTOR

Lycopene is a powerful antioxidant linked to reduced risk of certain cancers and heart disease. Cooking dramatically increases its bioavailability:

- Raw tomatoes: About 3% of lycopene absorbed
- Cooked tomatoes: Up to 30% of lycopene absorbed

Cooking also changes the form of lycopene from trans-lycopene to cis lycopene, which is more easily absorbed by the body.

TOMATO COOKING STUDY

A study published in the Journal of Agriculture and Food Chemistry found that cooking tomatoes in olive oil significantly increased the absorption of lycopene. The fat in the oil helps the body absorb this fat-soluble nutrient.

OTHER NUTRIENT CHANGES

- Vitamin A levels can increase with cooking

- Some B vitamins may decrease
- Mineral content generally stays stable

FLAVOR CHEMISTRY

Cooking tomatoes creates new flavor compounds through the Maillard reaction and caramelization. Over 400 volatile compounds contribute to the complex flavor of cooked tomatoes!

REAL-LIFE STORY: THE CURIOUS CHEF

Let's revisit Alex, our software engineer turned kitchen scientist:

"After learning about the Maillard reaction, I became obsessed with the chemistry of cooking," Alex says. "I started experimenting with different cooking methods and how they affected both flavor and nutrition."

ALEX'S COOKING EXPERIMENTS

1. **Tomato Sauce Test:** "I made two batches of tomato sauce - one with olive oil and one without. The one with oil not only tasted better but knowing it increased lycopene absorption made me feel like a kitchen genius!"
2. **Broccoli Challenge:** "I steamed broccoli instead of boiling it and was amazed at how much more flavor it had. Plus, I learned it retained more nutrients this way."
3. **Meat Marinade Magic:** "After reading about how marinades can reduce the formation of harmful

compounds when grilling, I created my own antioxidant-rich marinade with herbs and citrus. The meat tasted incredible, and I felt good knowing it was a healthier way to grill."

Alex's journey shows how understanding food chemistry can transform everyday cooking into a fascinating exploration of science, nutrition, and flavor.

BRINGING IT ALL TOGETHER: THE CHEMISTRY OF A HOME-COOKED MEAL

Let's analyze our simple meal of grilled chicken, roasted vegetables, and a salad from a chemical perspective:

1. GRILLED CHICKEN

- Maillard reaction between amino acids and sugars creates a flavorful brown crust
- Proteins denature, changing the texture of the meat
- Marinading in acidic ingredients (like lemon juice) can help prevent the formation of HCAs
- Grilling allows fat to drip away, reducing overall fat content

Chemical Tip: Adding rosemary to your marinade can reduce HCA formation by up to 92%, thanks to its antioxidant compounds.

2. ROASTED VEGETABLES

- Caramelization of sugars enhances sweetness
- Heat breaks down cell walls, softening the vegetables
- Some nutrients may be lost, but others (like lycopene in tomatoes) become more available
- Olive oil used in roasting helps absorb fat-soluble vitamins

Vegetable Roasting Study: Research published in the Journal of Food Science found that roasting increased the total antioxidant ability of several vegetables, including carrots, broccoli, and Brussels sprouts.

3. FRESH SALAD

- Raw vegetables keep heat-sensitive vitamins
- Adding oil-based dressing helps absorb fat-soluble vitamins from the vegetables
- Cutting the vegetables releases enzymes that can lead to browning (like in apples)
- Some vegetables, like garlic, release beneficial compounds when cut

Salad Science: Drizzling lemon juice on cut apples can prevent browning by inhibiting the enzyme polyphenol oxidase.

THE FUTURE OF FOOD CHEMISTRY

As our understanding of food chemistry grows, new techniques and technologies are appearing:

1. SOUS VIDE COOKING

- Food is sealed in plastic bags and cooked in temperature-controlled water
- Preserves nutrients and creates unique textures
- Allows precise control over doneness and flavor development

2. MOLECULAR GASTRONOMY

- Applies scientific principles to create new textures and flavors
- Examples: fruit caviar, edible foams, flavor-changing foods
- Challenges traditional concepts of how food should look and taste

3. 3D FOOD PRINTING

- Creates custom-shaped foods with precise nutritional profiles
- Potential to create personalized meals based on individual nutritional needs
- Could revolutionize food production and consumption

4. NUTRIGENOMICS

- Studies how different foods interact with our genes

- Could lead to personalized diet recommendations based on genetic profiles
- May help prevent diet-related diseases

5. SMART KITCHENS

- AI-powered appliances that can analyze food and suggest the best cooking methods
- Sensors that detect doneness and nutrient content in real-time
- Could help home cooks maximize both flavor and nutrition

WRAPPING IT UP

We've journeyed through the complex worlds of organic food, GMOs, and cooking chemistry. Here are the key takeaways:

1. Organic food isn't necessarily pesticide-free or more nutritious, but it can support more sustainable farming practices and reduce exposure to synthetic pesticides.
2. GMOs are generally considered safe to eat and can have environmental benefits, but their long-term impacts are still being studied. The debate involves not just science, but also economic, cultural, and ethical considerations.
3. Cooking methods can significantly change the nutritional content and chemical composition of our food. Understanding these changes can help us make choices that balance nutrition, safety, and flavor.

4. The future of food science is exciting, with modern technologies promising to revolutionize how we produce, prepare, and understand our food.

Remember, food chemistry isn't just for scientists - it's something we all engage with every time we cook a meal or choose what to eat. By understanding the chemistry of our food, we can make informed choices that balance nutrition, flavor, environmental impact, and personal values.

So, the next time you're in the kitchen or the grocery store, think about the amazing chemical transformations happening in your food. Experiment with different cooking methods, try new ingredients, and don't be afraid to get a little scientific in your culinary adventures. After all, every kitchen is a laboratory, and every meal is an experiment in delicious chemistry!

PART VI
CONSUMER INFORMATION AND REGULATION

Hey there, food detectives! We're about to embark on a journey through the world of food labels, regulations, and emerging concerns in our food supply. Grab your magnifying glass and let's dive in!

CHAPTER NINE
UNDERSTANDING FOOD LABELS AND CHEMICAL INGREDIENTS

Have you ever felt like you need a chemistry degree to understand food labels? Don't worry, we've got your back! Let's break down what all those mysterious ingredients really mean.

ANATOMY OF A FOOD LABEL

1. NUTRITION FACTS PANEL

- Shows calories, nutrients, and serving sizes
- Updated in 2020 to reflect current nutrition science

2. INGREDIENTS LIST

- Listed in descending order by weight
- Common allergens must be clearly labeled

3. HEALTH CLAIMS

- Regulated statements about nutrient content or health benefits

- **Example:** "High in fiber" or "May reduce risk of heart disease"

4. ORGANIC SEAL

- USDA certified organic products
- Means 95% or more organic ingredients

DECODING CHEMICAL INGREDIENTS

Let's demystify some common chemical-sounding ingredients:

1. ASCORBIC ACID

- Fancy name for Vitamin C
- Used as an antioxidant and preservative

2. TOCOPHEROLS

- Forms of Vitamin E
- Natural preservatives that prevent rancidity

3. LECITHIN

- Usually from soybeans or eggs
- Emulsifier that helps mix oil and water

4. CARRAGEENAN

- Extracted from seaweed
- Thickener and stabilizer in dairy products

5. MALTODEXTRIN

- Made from corn, wheat, or potato starch
- Used as a thickener or filler

6. SODIUM BENZOATE

- Preservative that prevents growth of bacteria and fungi
- Found in acidic foods like salad dressings and sodas

7. BUTYLATED HYDROXYANISOLE (BHA) AND BUTYLATED HYDROXYTOLUENE (BHT)

- Synthetic antioxidants
- Preserve fats and oils, but controversial due to potential health risks

Medical Backup: A study published in the International Journal of Toxicology in 2020 reviewed the safety of BHA and BHT. While generally recognized as safe (GRAS) by the FDA, some animal studies suggest potential carcinogenic effects at high doses. The study concluded that more research is needed to fully understand long-term effects in humans.

NATURAL VS. ARTIFICIAL FLAVORS

Ever wondered about the difference between "natural" and "artificial" flavors? Here's the scoop:

NATURAL FLAVORS

- Derived from plant or animal sources
- Can be highly processed

- Not necessarily healthier than artificial flavors

ARTIFICIAL FLAVORS

- Created in a lab to mimic natural flavors
- Often more stable and consistent than natural flavors
- May be cheaper to produce

Fun Fact: Both natural and artificial banana flavoring contain the chemical isoamyl acetate.

The natural version is extracted from bananas, while the artificial version is synthesized in a lab, but they're chemically identical!

TIPS FOR READING FOOD LABELS

1. **Look at serving sizes:** They might be smaller than you think!
2. **Check for added sugars:** The new label makes this easier to spot.
3. **Be wary of health claims:** "Low-fat" doesn't always mean healthy.
4. **Look for whole food ingredients:** The shorter the ingredient list, often the better.
5. **Know your additives:** Some are harmless, others you might want to avoid.

CHAPTER TEN
REGULATORY FRAMEWORK FOR FOOD CHEMICALS

Ever wonder who's making sure our food is safe? Let's explore the world of food regulation!

KEY PLAYERS IN FOOD REGULATION

1. FDA (FOOD AND DRUG ADMINISTRATION)

- Regulates most food products in the U.S.
- Ensures food additives and chemicals are safe

2. USDA (UNITED STATES DEPARTMENT OF AGRICULTURE)

- Regulates meat, poultry, and eggs
- Oversees organic certification

3. EPA (ENVIRONMENTAL PROTECTION AGENCY)

- Sets limits for pesticide residues on food

4. STATE AND LOCAL HEALTH DEPARTMENTS

- Inspect restaurants and food manufacturing facilities
- Enforce food safety regulations locally

HOW FOOD ADDITIVES ARE REGULATED

Before a new food additive hits the market, it goes through a rigorous process:

1. SAFETY TESTING

- Animal studies to decide potential health effects
- Establishment of Acceptable Daily Intake (ADI)

2. PETITION TO FDA

- Company sends data showing additive is safe
- FDA reviews data and may request more studies

3. FDA APPROVAL

- If believed safe, FDA issues a regulation for its use
- Specifies allowed uses and maximum levels

4. GRAS LIST

- Some substances are "Generally Recognized as Safe"
- Can be used without formal FDA approval process

Medical Backup: A 2019 study in Comprehensive Reviews in Food Science and Food Safety examined the FDA's regulatory process for food additives. The study found that while the process is generally robust, there are concerns about the lack of systematic post-market surveillance for long-term effects of additives.

INTERNATIONAL REGULATIONS

Food regulation isn't just a U.S. thing. Let's look at some international players:

1. CODEX ALIMENTARIUS COMMISSION

- Joint WHO and FAO body setting international food standards
- Aims to ensure fair trade practices and protect consumer health

2. EUROPEAN FOOD SAFETY AUTHORITY (EFSA)

- Scientific body providing risk assessment for EU food policies
- Often stricter than U.S. regulations

3. FOOD STANDARDS AUSTRALIA NEW ZEALAND (FSANZ)

- Develops food standards for Australia and New Zealand
- Known for stringent labeling requirements

CHALLENGES IN FOOD REGULATION

Regulating our food supply isn't easy. Here are some challenges:

1. EMERGING TECHNOLOGIES

- Nanoparticles in food packaging
- Gene-edited crops

- Lab-grown meat

2. GLOBAL FOOD SUPPLY

- Different standards in different countries
- Tracking ingredients from multiple sources

3. "CLEAN LABEL" TREND

- Consumer demand for fewer additives
- Challenge of supporting food safety and shelf life

4. SCIENTIFIC UNCERTAINTY

- Long-term effects of some additives still unknown
- Difficulty in studying cumulative effects of multiple additives

SHORT STORY: THE CURIOUS CASE OF RED DYE #2

In the 1970s, Mary Thompson was a young FDA scientist. She noticed something odd in the data on Red Dye #2, a popular food coloring.

"The more I looked, the more concerned I became," Mary recalls. "The animal studies showed potential cancer risks, but the dye was everywhere - in cereals, candies, even medications."

Mary raised the alarm but faced pushback from industry and even some colleagues. "It was a tough battle," she says. "But I knew we had to put public health first."

After years of debate and other studies, the FDA banned Red Dye #2 in 1976. Mary's persistence had paid off, protecting countless consumers.

"It taught me the importance of rigorous science and standing up for what's right," Mary reflects. "Food safety isn't just about regulations - it's about real people's lives."

CHAPTER ELEVEN
EMERGING CONCERNS: NEW CHEMICALS IN OUR FOOD SUPPLY

Just when you thought you had food chemicals figured out, new ones pop up! Let's explore some emerging concerns in our food supply.

MICROPLASTICS: TINY PLASTICS, BIG WORRIES

WHAT ARE THEY?
- Plastic particles less than 5mm (about 0.2 in) in size
- Can come from breakdown of larger plastics or be intentionally manufactured

WHERE ARE THEY FOUND?
- Seafood (especially shellfish)
- Sea salt
- Bottled water

POTENTIAL HEALTH CONCERNS
- May absorb and concentrate environmental pollutants

- Could potentially enter human tissues
- Long-term health effects still unknown

Medical Backup: A 2019 study in the journal Environmental Science & Technology estimated that the average American ingests 39,000 to 52,000 microplastic particles per year. The health implications of this ingestion are still being studied.

PER- AND POLYFLUOROALKYL SUBSTANCES (PFAS): THE "FOREVER CHEMICALS"

WHAT ARE THEY?

- Man-made chemicals used in non-stick cookware, food packaging, and firefighting foam
- Extremely persistent in the environment and human body

WHERE ARE THEY FOUND?

- Some fast-food packaging
- Non-stick cookware
- Contaminated water supplies

POTENTIAL HEALTH CONCERNS

- Linked to cancer, thyroid disease, and immune system problems
- Can accumulate in the body over time

- Difficult to avoid due to widespread environmental contamination

Medical Backup: A 2020 review in the journal Environmental Health Perspectives summarized the health effects of PFAS exposure. The review found consistent evidence linking PFAS exposure to altered immune and thyroid function, liver enzyme levels, and fetal growth.

NANOPARTICLES: THE TINY TITANS

WHAT ARE THEY?

- Extremely small particles (1-100 nanometers)
- Used in food for various purposes like improving texture or adding nutrients

WHERE ARE THEY FOUND?

- Some food packaging (to improve barrier properties)
- Certain food products (e.g., titanium dioxide as a whitening agent)
- Supplements (e.g., nano-sized minerals for better absorption)

POTENTIAL HEALTH CONCERNS

- May behave differently in the body compared to larger particles
- Could potentially cross cellular barriers
- Long-term health effects not yet fully understood

Medical Backup: A 2018 study in the journal Small found that ingested titanium dioxide nanoparticles could accumulate in the liver and trigger immune responses in mice. However, more research is needed to understand the implications for human health.

EMERGING PLANT-BASED MEAT ALTERNATIVES

As plant-based meats gain popularity, new ingredients and processing methods are being used:

- **Soy leghemoglobin:** Genetically engineered yeast produces this plant-based heme protein
- **Methylcellulose:** Used as a binder in many plant-based meats
- **Various protein isolates:** From peas, mung beans, and other plants

POTENTIAL CONCERNS

- Novel ingredients with limited long-term consumption data
- Highly processed nature of some products
- Potential allergenicity of new plant protein sources

Medical Backup: A 2020 review in the journal Foods examined the nutritional and health aspects of plant-based meat alternatives. While generally considered safe, the review noted the need for more research on long-term health effects and potential allergenicity of novel ingredients.

TACKLING EMERGING FOOD CHEMICAL CONCERNS

So, what can we do about these new chemical concerns? Here are some strategies:

1. STAY INFORMED

- Follow reputable science news sources
- Be critical of sensationalized headlines

2. SUPPORT RESEARCH

- Advocate for more funding for food safety research
- Take part in citizen science projects when available

3. MAKE SMART CONSUMER CHOICES

- Choose fresh, whole foods when possible
- Reduce use of heavily processed foods and single-use plastics

4. ADVOCATE FOR BETTER REGULATIONS

- Support organizations working for stricter chemical regulations
- Contact your representatives about food safety concerns

5. PRACTICE SAFE FOOD HANDLING

- Avoid heating food in plastic containers

- Use stainless steel or glass water bottles instead of plastic

THE FUTURE OF FOOD SAFETY

As our understanding of food chemistry grows, so does our ability to ensure a safe food supply. Here are some exciting developments:

1. BLOCKCHAIN TECHNOLOGY

- Could improve traceability in the food supply chain
- Help quickly name sources of contamination

2. ARTIFICIAL INTELLIGENCE

- Predict potential safety issues in new food additives
- Analyze large datasets to find emerging risks

3. PRECISION FERMENTATION

- Produce food ingredients more efficiently and sustainably
- Could reduce reliance on traditional agriculture and its associated chemical inputs

4. EDIBLE FOOD PACKAGING

- Reduce plastic waste
- Made from natural, biodegradable materials

5. PERSONALIZED NUTRITION

- Tailor diets based on individual genetic profiles

- Could help avoid foods or additives that may cause personal sensitivities

CHAPTER TWELVE
PERSONALIZED NUTRITION - THE FUTURE OF EATING

Welcome to the forefront of nutritional science! In this chapter, we'll explore the exciting world of personalized nutrition. Gone are the days of one-size-fits-all dietary advice. We're entering an era where your genes, your microbiome, and advanced technologies all play a role in deciding the best diet for you. Let's dive in!

1. GENETIC FACTORS IN NUTRITION

Our genes play a crucial role in how we process and use nutrients. This field of study, known as nutrigenomics, is revolutionizing our understanding of the relationship between diet and health.

1.1 - NUTRIGENOMICS: THE BASICS

Nutrigenomics examines how different foods interact with specific genes to influence health outcomes. It's based on the idea that what we eat can affect how our genes express themselves (gene expression), and conversely, our genetic makeup can influence how we respond to different nutrients.

Key Concepts in Nutrigenomics:

- **Single Nucleotide Polymorphisms (SNPs):** These are small variations in DNA sequences that can affect how we metabolize nutrients.
- **Epigenetics:** This refers to changes in gene expression that don't involve changes to the underlying DNA sequence. Diet can influence epigenetic changes.

1.2 - GENETIC VARIATIONS AND NUTRIENT METABOLISM

Let's look at some specific examples of how genetic variations can affect nutrient needs:

a) Lactose Intolerance:

The LCT gene provides instructions for making the enzyme lactase, which breaks down lactose in dairy products. Variations in this gene can lead to lactose intolerance.

b) Caffeine Metabolism:

The CYP1A2 gene affects how quickly we metabolize caffeine. Some people are "fast metabolizers" and can drink coffee late in the day without sleep issues, while "slow metabolizers" might experience insomnia from afternoon coffee.

c) Vitamin Needs:

- **MTHFR Gene:** Variations in this gene can affect how efficiently the body processes folate and may increase the need for folate supplementation.

- **VDR Gene:** Variations here can influence vitamin D metabolism and may affect best vitamin D intake levels.

d) Fat Metabolism:

The APOA2 gene influences how the body responds to saturated fat intake. Some variations may increase the risk of obesity when consuming a high saturated fat diet.

1.3 - IMPLICATIONS FOR PERSONALIZED NUTRITION

Understanding an individual's genetic profile can help tailor dietary recommendations. For example:

- Someone with a genetic predisposition to high cholesterol might receive help from a diet lower in saturated fats.
- An individual with certain MTHFR gene variations might need to focus on consuming more folate-rich foods or consider supplementation.
- People with genetic variations affecting caffeine metabolism might need to be more mindful of their coffee intake, especially later in the day.

However, it's crucial to remember that genes are just one piece of the puzzle. Environmental factors, lifestyle choices, and other variables also play significant roles in health outcomes.

2. THE MICROBIOME AND NUTRITION

The human microbiome - the trillions of microorganisms living in and on our bodies - is appearing as a key player in nutrition and overall health. Let's explore how these tiny residents influence our dietary needs and responses.

2.1 - UNDERSTANDING THE GUT MICROBIOME

The gut microbiome consists of billions of bacteria, fungi, and other microorganisms living in our digestive tract. These microbes play crucial roles in:

- Digestion and nutrient absorption
- Immune system function
- Production of certain vitamins (like K and B12)
- Influencing mood and mental health
- Metabolism and weight regulation

2.2 - HOW DIET SHAPES THE MICROBIOME

What we eat has a profound impact on our gut microbiome:

- Fiber-rich foods feed beneficial bacteria
- Fermented foods can introduce beneficial probiotics
- Highly processed foods may promote less diverse, less healthy microbial communities

Interestingly, the relationship is bidirectional - while our diet shapes our microbiome, our microbiome also influences our food cravings and how we metabolize nutrients.

2.3 - MICROBIOME DIVERSITY AND HEALTH

A diverse microbiome is generally associated with better health outcomes. Factors that can reduce microbiome diversity include:

- Overuse of antibiotics
- Lack of dietary fiber
- Chronic stress
- Lack of sleep

2.4 - THE MICROBIOME AND PERSONALIZED NUTRITION

Emerging research suggests that individual differences in gut microbiome composition can influence how we respond to different diets. For example:

- Some people may extract more calories from certain foods due to their specific microbial makeup.
- The effectiveness of probiotics can vary based on an individual's existing microbiome.
- Certain microbial profiles might predispose individuals to food intolerances or allergies.

2.5 - MICROBIOME TESTING AND DIETARY INTERVENTIONS

Microbiome testing is becoming increasingly available to consumers. These tests analyze stool samples to provide insights into microbial diversity and abundance. Based on these results, personalized dietary recommendations might include: - Increasing intake of specific prebiotic fibers to feed beneficial bacteria

- Adding fermented foods or probiotic supplements
- Avoiding foods that may promote growth of less beneficial microbes

However, it's important to note that microbiome science is still evolving, and interpreting test results for dietary recommendations is complex.

3. EMERGING TECHNOLOGIES IN PERSONALIZED NUTRITION

Exciting technological advances are making personalized nutrition more accessible and precise than ever before. Let's explore some of these innovative tools and approaches.

3.1 - GENETIC TESTING FOR NUTRITION

Direct-to-consumer genetic tests now offer nutrition-related insights: - Companies like 23andMe and Ancestry provide some nutrition-related genetic information.

- Specialized nutrigenomics companies offer more detailed dietary advice based on genetic profiles.
- Limitations:
- The science is still evolving, and recommendations may change as we learn more.
- Genetic factors are just one aspect of health; lifestyle and environmental factors also play crucial roles.

3.2 - MICROBIOME TESTING AND ANALYSIS

As mentioned earlier, microbiome testing is becoming more widely available:

- Companies like Viome and uBiome offer at-home microbiome testing kits.
- These tests can provide insights into microbial diversity and potential imbalances.

Future Directions:
- More precise identification of beneficial and harmful microbes
- Better understanding of how to manipulate the microbiome through diet

3.3 - ARTIFICIAL INTELLIGENCE AND MACHINE LEARNING

AI and machine learning are being applied to nutrition in several ways:

- Analyzing large datasets to show patterns between diet, genetics, and health outcomes
- Developing more correct predictive models for personalized dietary recommendations
- Creating chatbots and virtual nutritionists for real-time dietary advice

Example: The startup Habit uses AI to analyze genetic, metabolic, and microbiome data to create personalized nutrition plans.

3.4 - WEARABLE TECHNOLOGY AND APPS

Wearable devices and smartphone apps are increasingly being used for nutrition tracking and guidance:

- Continuous glucose monitors can provide real-time data on how different foods affect blood sugar levels.
- Apps like MyFitnessPal and Chronometer allow detailed tracking of nutrient intake.
- Some smartwatches can now estimate hydration levels.

Future Possibilities:

- Wearables that can detect nutrient deficiencies in real-time
- Apps that use image recognition to automatically log food intake and provide nutritional information

3.5 - METABOLOMICS AND PERSONALIZED NUTRITION

Metabolomics - the study of small molecules (metabolites) in biological samples - is offering new insights into individual nutritional needs:

- Blood, urine, or saliva samples can be analyzed to create a metabolic profile.
- This profile can reveal how an individual is processing different nutrients and show potential deficiencies or imbalances.

Potential Applications:

- More precise vitamin and mineral supplementation recommendations
- Identification of food intolerances or sensitivities
- Tailoring macronutrient ratios to individual metabolic needs

3.6 - 3D PRINTED FOOD AND PERSONALIZED MEALS

While still in first stages, 3D food printing technology holds promise for personalized nutrition:

- Ability to create foods with precise nutrient compositions
- Potential to make personalized meals based on individual nutritional needs
- Could be especially beneficial for people with specific dietary restrictions or nutritional requirements

4. CHALLENGES AND ETHICAL CONSIDERATIONS

While personalized nutrition offers exciting possibilities, it also comes with challenges and ethical considerations:

4.1 - DATA PRIVACY AND SECURITY

Genetic and health data are sensitive and need robust protection.

Questions about who owns and has access to this data (individuals, companies, healthcare providers, insurers?)

4.2 - INTERPRETATION AND IMPLEMENTATION OF RESULTS

- Translating complex genetic and microbiome data into actionable dietary advice is challenging.
- Risk of oversimplification or misinterpretation of results
  4.3 Access and Equity
- Many personalized nutrition technologies are expensive and not widely accessible.
- Could lead to growing health disparities if only available to wealthier individuals.

4.3 - REGULATORY CHALLENGES

- How should direct-to-consumer genetic tests and personalized nutrition services be regulated?
- Ensuring accuracy of claims and preventing misleading information

4.4 - PSYCHOLOGICAL IMPACT

- How does knowing one's genetic predispositions affect behavior and mental health?
- Potential for increased anxiety or obsessive behavior around food choices

5. THE FUTURE OF PERSONALIZED NUTRITION

As we look to the future, several trends and possibilities appear:

5.1 - INTEGRATION WITH HEALTHCARE SYSTEMS

- Personalized nutrition could become a standard part of preventive healthcare.
- Potential for integration with electronic health records and collaboration with healthcare providers

5.2 - MORE SOPHISTICATED AI AND PREDICTIVE MODELS

- As we gather more data, AI models will become more exact in predicting individual responses to different diets.
- Could lead to highly precise, dynamic dietary recommendations that change based on real-time health data

5.3 - PERSONALIZED SUPPLEMENTS AND FUNCTIONAL FOODS

- Development of supplements tailored to individual genetic and microbiome profiles
- Creation of functional foods designed for specific genetic types or health conditions

5.4 - EDUCATION AND LITERACY

- Growing need for nutrition professionals trained in genetics and personalized nutrition
- Importance of improving public understanding of personalized nutrition concepts

SUMMARY

Personalized nutrition stands for a change in thinking in how we think about diet and health. By considering individual genetic factors, microbiome composition, and using advanced technologies, we're moving towards a future where dietary advice can be tailored with unprecedented precision.

However, it's crucial to remember that personalized nutrition is not a magic bullet. The foundations of a healthy diet - plenty of vegetables, fruits, whole grains, and lean proteins - remain important for most people. Personalized approaches should be seen to fine-tune these basics, not replace them entirely.

As this field continues to evolve, it holds the promise of more effective nutrition strategies for preventing and managing chronic diseases, perfecting performance, and promoting overall well-being. The key will be balancing the exciting possibilities with careful consideration of the ethical, practical, and scientific challenges involved.

Whether you're a nutrition enthusiast, a healthcare professional, or simply someone interested in improving your diet, staying informed about developments in personalized nutrition will be increasingly important in the years to come. Who knows - the perfect diet for you might be just a genetic test or microbiome analysis away!

WRAP UP

Whew! We've covered a lot of ground in our exploration of food labels, regulations, and emerging chemical concerns. Here are the key takeaways:

1. Understanding food labels empowers you to make informed choices about what you eat.
2. Food regulations aim to keep our food safe, but it's a complex and ever-evolving process.
3. New chemicals in our food supply present both opportunities and challenges, requiring ongoing research and vigilance.

Remember, knowledge is power when it comes to food safety. By staying informed and making conscious choices, you can navigate the complex world of food chemicals with confidence.

The next time your grocery shopping or preparing a meal, think about what you've learned. You're now equipped with the knowledge to be a savvy food consumer and an advocate for a safer, healthier food system. Happy eating, food detectives!

Thank you for joining us on this fascinating journey through the world of food chemistry! We've covered a lot of ground, and your dedication to understanding what's really on your plate is truly commendable. Let's take a moment to recap the key points we've explored together.

We began our adventure by diving into the basics of food chemistry, learning about the building blocks that make up our food. We discovered that everything we eat is made up of chemicals - some naturally occurring, others added during processing. Remember, the word "chemical" isn't inherently scary; it's simply a term for the substances that make up our world, including our food.

We explored the main components of food:

- **Carbohydrates:** Our body's primary energy source
- **Proteins:** The building blocks for our tissues and organs
- **Fats:** Essential for nutrient absorption and energy storage
- **Vitamins and Minerals:** Crucial micronutrients for various bodily functions
- **Water:** The often overlooked but vital part of our diet

Next, we delved into the world of food additives and preservatives. We learned that while some additives are synthetic and potentially controversial, others are natural and have been used for centuries. We discovered the roles of various additives:

- Preservatives to extend shelf life
- Colorants to make food more visually appealing
- Flavor enhancers to improve taste
- Texturizers to create the right mouthfeel

We also explored the impact of pesticides and herbicides on our food supply. While these chemicals help increase crop yields and protect against pests, we learned about potential health and environmental concerns associated with their use. We discussed the "Dirty Dozen" and "Clean Fifteen" lists, which can guide us in making choices about when to prioritize organic produce.

Our journey then took us into the kitchen, where we explored how different cooking methods affect the chemistry of our food. We learned that cooking isn't just about making food taste good - it can also significantly change nutritional content. We discovered:

- How boiling can lead to nutrient loss in water-soluble vitamins
- The Maillard reaction that gives grilled and fried foods their distinctive flavors
- How pressure cooking can make nutrients more bioavailable
- The pros and cons of raw food diets

We tackled the sometimes-controversial topic of Genetically Modified Organisms (GMOs). We learned about the science behind genetic modification, common GMO crops, and the ongoing debate about their safety and environmental impact. We discovered that while scientific consensus suggests GMOs are safe to eat, there are still concerns about their broader ecological effects.

Our exploration of organic food helped us understand what the "organic" label really means. We debunked some common myths and learned about the potential benefits and limitations of organic farming practices. We discovered that while organic food can reduce exposure to synthetic pesticides, it's not necessarily more nutritious than conventional produce.

We then turned our attention to food packaging and the potential for chemical migration into our food. We learned about different packaging materials and their pros and cons:

- Plastics and concerns about chemicals like BPA
- The benefits and limitations of glass packaging
- The potential for chemical migration from metal cans
- Appearing sustainable packaging options

Our journey also included a deep dive into food labeling and regulations. We learned how to decode ingredient lists, understand nutrition facts panels, and critically evaluate health claims on food packages. We explored the roles of various regulatory agencies like the FDA, USDA, and EPA in ensuring food safety.

Finally, we looked at emerging concerns in our food supply, including:

- Microplastics finding their way into our food chain
- "Forever chemicals" like PFAS in food packaging
- The potential impacts of nanoparticles in food
- Novel ingredients in plant-based meat alternatives

Throughout our exploration, we've seen how food chemistry intersects with health, environmental concerns, and even social and ethical issues. We've learned that while our food supply is generally safe, there's always room for improvement and ongoing research.

KEY TAKEAWAYS

1. **Knowledge is power:** Understanding food chemistry empowers us to make informed choices about what we eat.
2. **Balance is key:** While it's important to be aware of potential risks, it's equally important not to become overly fearful. A balanced, varied diet is generally the best approach.
3. **Science is evolving:** Our understanding of food chemistry is constantly improving. Stay curious and keep learning!
4. **Personal choice matters:** What we choose to eat can affect not just our health, but also the environment and food systems at large.

5. **Advocacy is important:** As consumers, we have the power to influence food policies and industry practices through our choices and by making our voices heard.

As we conclude our journey, remember that every meal is an opportunity to apply what you've learned. Whether you're deciphering a food label, choosing between organic and conventional produce, or deciding how to cook your dinner, you now have the tools to make informed decisions.

Thank you for your curiosity and dedication throughout this exploration of food chemistry. Your commitment to understanding what's really in your food is not just personally beneficial - it contributes to a more informed society that can push for better, safer, and more sustainable food systems.

Remember, the journey doesn't end here. The world of food chemistry is constantly evolving, with new discoveries and challenges appearing all the time. Stay curious, keep asking questions, and never stop exploring the fascinating science behind what's on your plate.

Here's to many more delicious and informed meals ahead!

Bon appétit, Food Explorer!

Thanks for reading
Kevin

GLOSSARY OF TERMS

A

Additives: Substances added to food to preserve flavor, enhance taste or appearance, or serve other purposes.

Amino Acids: The building blocks of proteins, essential for various bodily functions.

Antioxidants: Compounds that can prevent or slow damage to cells caused by free radicals.

Artificial Flavors: Flavoring substances not derived from natural sources, created in a laboratory.

B

Bioaccumulation: The gradual accumulation of substances, such as pesticides or other chemicals, in an organism.

Bioavailability: The extent to which a nutrient can be absorbed and used by the body.

Bisphenol A (BPA): A chemical used in some plastics and resins, particularly in food packaging.

C

Carbohydrates: One of the main types of nutrients, primarily responsible for providing energy to the body.

Caramelization: The browning of sugar when heated, resulting in complex flavors.

Carcinogens: Substances capable of causing cancer in living tissue.

Certified Organic: Foods produced according to strict USDA organic standards.

Chemical Migration: The transfer of chemical substances from packaging materials into food.

Codex Alimentarius: A collection of internationally recognized standards, codes of practice, guidelines, and recommendations relating to food, food production, and food safety.

Complex Carbohydrates: Carbohydrates composed of long chains of sugar molecules, typically found in starchy foods.

D

Denaturation: The alteration of a protein's structure by heat, acid, or other means, often changing its properties.

Dietary Fiber: The indigestible part of plant foods that plays a vital role in digestion.

Dioxins: A group of highly toxic chemical compounds that are byproducts of various industrial processes.

E

Emulsifiers: Substances that help mix liquids that normally don't combine, incompatible.

Enzymes: Proteins that act as biological catalysts, speeding up chemical reactions in the body.

Essential Amino Acids: Amino acids that cannot be produced by the body and must be obtained from food.

F

Fermentation: A metabolic process that converts sugar to acids, gases, or alcohol, used in the production of many foods.

Food Additives: Substances added to food to preserve flavor, enhance taste or appearance, or serve other purposes.

Free Radicals: Unstable molecules that can damage cells, potentially leading to various diseases.

G

Genetically Modified Organism (GMO): An organism whose genetic material has been altered using genetic engineering techniques.

Gelatinization: The process where starch granules absorb water and swell when heated, often used in cooking.

Generally Recognized as Safe (GRAS): A designation by the FDA for food additives considered safe by experts.

Gluten: A protein found in wheat, barley, and rye that gives dough its elastic texture.

H

Hydrogenation: The process of adding hydrogen to liquid oils to make them more solid, often creating trans fats.

I

Insoluble Fiber: A type of dietary fiber that does not dissolve in water, aiding in digestion.

Irradiation: The process of exposing food to ionizing radiation to kill bacteria, viruses, or insects.

J

Junk Food: Foods that are high in calories but low in nutritional value.

K

Ketones: Compounds produced when fat is broken down for energy, central to low-carb diets.

L

Lactose: The sugar found in milk and dairy products.

Lipids: A broad group of naturally occurring molecules including fats, waxes, and vitamins.

Lycopene: A bright red carotenoid pigment found in tomatoes and other red fruits and vegetables.

M

Macronutrients: Nutrients that the body needs in substantial amounts: protein, carbohydrates, and fats.

Maillard Reaction: A chemical reaction between amino acids and reducing sugars that gives browned foods their distinctive flavor.

Micronutrients: Nutrients required in small quantities to ensure normal metabolism, including vitamins and minerals.

Microplastics: Tiny pieces of plastic less than 5mm (about 0.2 in) in length, which can contaminate food and water.

Mineral: Inorganic nutrients essential for various bodily functions.

Monosodium Glutamate (MSG): A flavor enhancer commonly added to Chinese food, canned vegetables, soups, and processed meats.

N

Natural Flavors: Flavoring substances derived from plant or animal sources.

Nitrates/Nitrites: Compounds used as preservatives in processed meats.

Nutrient Density: The ratio of nutrients to calories in a food.

O

Omega-3 Fatty Acids: A type of polyunsaturated fat important for heart and brain health.

Organic: Produced without the use of synthetic pesticides, herbicides, or fertilizers.

Oxidation: A chemical reaction involving the loss of electrons, often leading to food spoilage.

P

Pasteurization: The process of heating a food to a specific temperature for a set time to kill harmful bacteria.

Pesticides: Chemical substances used to kill pests, including insects, rodents, fungi and unwanted plants.

pH: A measure of how acidic or basic a substance is.

Phytochemicals: Compounds produced by plants that may have health benefits when consumed.

Preservatives: Additives that prevent or inhibit spoilage of food due to bacteria, fungi, or other microorganisms.

Probiotics: Live bacteria and yeasts that are good for health, especially the digestive system.

Proteins: Large, complex molecules that play many critical roles in the body, composed of amino acids.

Q

Quality Control: A system of supporting standards in manufactured products by testing a sample of the output against the specification.

R

Rancidity: The oxidation of fats and oils that can produce unpleasant odors or flavors in food.

Recommended Dietary Allowance (RDA): The average daily level of intake sufficient to meet the nutrient requirements of nearly all healthy people.

Refined Grains: Grains that have been milled, a process that removes the bran and germ.

S

Saturated Fat: A type of fat in which the fatty acid chains have no double bonds between carbon atoms.

Simple Carbohydrates: Carbohydrates composed of one or two sugar molecules, quickly absorbed by the body.

Soluble Fiber: A type of dietary fiber that dissolves in water, potentially helping to lower blood cholesterol and glucose levels.

Stabilizers: Additives used to keep the physical characteristics of food products.

T

Trans Fats: Unsaturated fats that have been hydrogenated, linked to increased risk of heart disease.

Trace Minerals: Minerals needed in tiny amounts for various bodily functions.

U

Umami: One of the five basic tastes, described as savory or meaty.

Unsaturated Fat: A fat or fatty acid in which there is at least one double bond within the fatty acid chain.

V

Vitamins: Organic compounds essential for normal growth and nutrition that are needed in small quantities in the diet.

Volatile Organic Compounds (VOCs): Organic chemicals that have a high vapor pressure at room temperature, some of which contribute to food flavors.

W

Whole Grains: Grains that have the entire grain kernel — the bran, germ, and endosperm.

X

Xanthan Gum: A polysaccharide used as a food additive and rheology modifier.

Y

Yeast: A single-celled microorganism used in baking and brewing for fermentation.

Z

Zein: A class of prolamine protein found in maize (corn), sometimes used in food packaging.

ADDITIONAL TERMS

Acidity Regulator: A food additive used to change or support ph.

Aflatoxins: Poisonous carcinogens produced by certain molds found in food.

Antioxidants: Substances that inhibit oxidation, often used as food preservatives.

Aspartame: An artificial non-saccharide sweetener used as a sugar substitute.

Bioengineered Foods: Foods that hold detectable genetic material that has been changed through certain lab techniques and cannot be created through conventional breeding or found in nature.

Calorie: A unit of energy, used to measure the energy content of foods.

Carotenoids: Pigments found in plants and algae, some of which can be converted to vitamin A.

Cellulose: An indigestible carbohydrate that is the main part of plant cell walls.

Cholesterol: A waxy substance found in all cells in your body, necessary for producing hormones, vitamin D, and substances that help you digest foods.

Collagen: The main structural protein found in skin and other connective tissues.

Dextrin: A group of low-molecular-weight carbohydrates produced by the hydrolysis of starch.

Emulsion: A mixture of two or more liquids that are normally immiscible.

Enrichment: The addition of specific nutrients to foods.

Enzyme: A substance produced by a living organism that acts as a catalyst to bring about a specific biochemical reaction.

Essential Fatty Acids: Fatty acids that are needed but cannot be synthesized by the human body.

Fermentation: The chemical breakdown of a substance by bacteria, yeasts, or other microorganisms.

Flavonoids: A class of plant and fungus secondary metabolites known for their health benefits.

Food Fortification: The process of adding micronutrients to food.

Free Sugars: Monosaccharides and disaccharides added to foods by the manufacturer, cook or consumer, plus sugars naturally present in honey, syrups and fruit juices.

Functional Foods: Foods that have a potentially positive effect on health beyond basic nutrition.

Glycemic Index: A measure of how quickly a food can elevate blood sugar levels.

High-Fructose Corn Syrup: A sweetener made from corn starch, used in many processed foods.

Homogenization: A process used to mix milk fat evenly throughout milk.

Hydrocolloids: Substances that form colloids mixed with water, often used as thickeners or stabilizers in food.

Hydrolysis: The chemical breakdown of a compound due to reaction with water.

Lacto-fermentation: A process in which lactic-acid producing bacteria break down sugars.

Lecithin: A fat that is essential in the cells of the body, often used as an emulsifier in food production.

Lipid Oxidation: The process in which fatty acids react with oxygen, leading to rancidity.

Maceration: Softening or breaking into pieces using a liquid.

Maltodextrin: A polysaccharide used as a food additive.

Microbiome: The microorganisms in a particular environment, including the body or a part of the body.

Nutraceuticals: Products derived from food sources that provide extra health benefits in addition to the basic nutritional value found in foods.

Nutrigenomics: The study of the effects of foods and food constituents on gene expression.

Osmosis: The movement of a solvent through a semipermeable membrane from a less concentrated solution into a more concentrated one.

Oxidative Stress: An imbalance between free radicals and antioxidants in your body.

Pectin: A soluble fiber found in fruits, used as a gelling agent in foods.

Phytosterols: Compounds found in plants that have a chemical structure like cholesterol.

Polyphenols: Micronutrients with antioxidant activity, widely found in fruits, vegetables, herbs and spices.

Prebiotics: Specialized plant fibers that act as food for the gut microbiome.

Proteolysis: The breakdown of proteins into smaller polypeptides or amino acids.

Rancidification: The process of complete or incomplete oxidation of fats and oils when exposed to air, light, or moisture.

Retrogradation: The reassociation of gelatinized starch molecules as they cool and age.

Saponification: A process that produces soap, usually from fats and lye.

Satiety: The feeling of fullness after eating

Simple Sugars: Monosaccharides and disaccharides, including glucose, fructose, and sucrose.

Starch: A complex carbohydrate consisting of many glucose units joined by glycosidic bonds.

Sterilization: The process of destroying all microorganisms in food for preservation.

Sulfites: Compounds used as preservatives in dried fruits, wines, and other foods.

Superfoods: Nutrient-rich foods considered to be especially beneficial for health and well-being.

Sweeteners (Artificial): Non-sugar substances used to sweeten food and drinks.

Synbiotics: Products that have both probiotics and prebiotics.

Tannins: Astringent biomolecules that bind to and precipitate proteins and various other organic compounds.

Texture Modifiers: Substances added to food to alter its mouthfeel or physical properties.

Thickeners: Substances used to increase the viscosity of a liquid without substantially changing its other properties.

Tocopherols: A class of organic chemical compounds, many of which have vitamin E activity.

Transgenic: Relating to an organism that holds genetic material into which DNA from an unrelated organism has been artificially introduced.

Trypsin Inhibitors: Substances that reduce the availability of trypsin, an enzyme involved in the digestion of protein.

Ultrapasteurization: A food processing technique that sterilizes liquid food by heating it above 137.8°C (280°F).

Umami: One of the five basic tastes, described as savory and characteristic of broths and cooked meats.

Unsaturated Fat: A fat or fatty acid in which there is at least one double bond within the fatty acid chain.

Vacuum Packaging: A method of packaging that removes air from the package prior to sealing.

Viscosity: A measure of a fluid's resistance to flow.

Vitamin Fortification: The process of adding vitamins to food products.

Water Activity: The measure of free water in food that is available for microbial growth.

Whole Food: Food that has been processed or refined as little as possible and is free from additives or other artificial substances.

Xenobiotics: Chemical substances found within an organism that are not naturally produced or expected to be present within the organism.

Yeast Extract: A food flavoring made from yeast, often used as a flavor enhancer.

Zein: A class of prolamine protein found in maize (corn), sometimes used in food packaging.

Zeaxanthin: A carotenoid alcohol found in many plants and some seafoods, important for eye health.

Zwitterion: A molecule with both positive and negative electrical charges, important in the behavior of amino acids and proteins in different pH environments.

This comprehensive glossary covers a wide range of terms related to food chemistry, nutrition, food processing, and food safety. It provides a solid foundation for understanding the complex world of food science and the many factors that influence what we eat. Remember, the field of food chemistry is constantly evolving, with added terms and concepts appearing as our understanding grows. Keeping this glossary handy can help you navigate food labels, understand nutritional information, and make informed decisions about your diet.

As you continue your journey into food chemistry, don't hesitate to expand this glossary with added terms you meet.

The world of food is vast and fascinating, and there's always more to learn! Thank You for reading.

Milton Keynes UK
Ingram Content Group UK Ltd.
UKHW022149190824
447134UK00016B/855